AF552468

BIOTECHNOLOGY OF ANIMAL TISSUES

BIOTECHNOLOGY OF ANIMAL TISSUES

By

Dr. P.R. Yadav
Lecturer
Department of Zoology
D.A.V. College
Muzaffarnagar (U.P.)

&

Dr. Rajiv Tyagi
Department of Zoology
M.M. College
Modi Nagar (U.P.)

DISCOVERY PUBLISHING HOUSE
NEW DELHI-110002

Published by:

DISCOVERY PUBLISHING HOUSE
4383/4B, Ansari Road, Darya Ganj
New Delhi-110 002 (India)
Phone : +91-11-23279245; 23253475; 43596065
E-mail : discoverybooksindia@gmail.com
discoverypublishinghouse@gmail.com
namitwasan9@gmail.com
web : www.discoverypublishinggroup.com

***First Published:* 2006**

***Reprinted:* 2022**

ISBN: 978-81-8356-084-9

Biotechnology of Animal Tissues

Printed at:
Infinity Imaging Systems
Delhi

Preface

The present title "Biotechnology of Animal Tissues" has been designed for all those who wish to know and use the principles and techniques of modern biotechnology, whether they are technicians needing further education graduate wishing to extend their knowledge base and potential, mature workers faced with changing work or a new career, managers unfamiliar with the new technology or those returning to work after the career break.

To cultivate animal cells *in vitro*, under pinning the understanding of animal cell biotechnology has been a developing capability. Cultivation of animal cells on an industrial scale enables the production of sufficient quantities of animal cell deprived materials to be of real value to society. This text is designed to describe the principles and techniques of culturing animal tissues, so that others may contribute to this important and extending area of endeavour.

In the preparation of this book large number of books and research papers have been consulted. So no authenticity is claimed.

The author wishes to express his deepest appreciation to the many people who have contributed in one way or the other to the preparation of this title.

The author expresses his gratitude to Mr. Wasan and staff of M/s Discovery Publishing House for their whole hearted co-operation in the publication of this book.

The author tried hard to be accurate and upto date in statement and realises the impossibility of completely avoiding errors therefore, the author will greatly appreciate having his attention called to any questionable statement.

Authors

Contents

1

Introduction

Cells, removed from animal tissue or whole animals, will continue to grow if supplied with nutrients and growth factors. This process is called cell cultures. It occurs *in vitro* ('in glass') as opposed to *in vivo* ('in life'). The culture process allows single cells to act as independent units much like any microorganism such as bacteria or fungi. The cells are capable of division by mitosis and the cell population can continue growth until limited by some parameter such as nutrient depletion. Animal cells selected for culture are maintained as independent units. Cultures normally contain cells of one type (e.g. *fibroblasts*).

The cells in the culture may be genetically identical (*homogeneous* population) or may show some genetic variation (*heterogeneous* population). A homogeneous population of cells derived from a single parental cell is called a clone. Therefore all cells within a clonal population are genetically identical. Cell culture is a widely used technique and is the main subject of this book. Cell culture is quite different from organ culture.

Organ culture means the maintenance of whole organs or fragments of tissue with the retention of a balanced relationship between the associated cell types as exists *in vivo*. This idea of the maintenance of tissue was important in the early development of culture techniques. However, it was soon realized that this was extremely difficult over long periods because of the differing growth potential of individual cell types within a tissue. Nowadays the culture of individual cells is the preferred technique because conditions can be controlled to allow some degree of consistency and reproducibility in cell growth. Although 'cell culture' is the most appropriate and logical term for this, it is still widely referred to as 'tissue culture'—a term that can be misleading, as it is also used to refer to organ culture.

Why Grow Animal Cells in Culture?

There are a number of applications for animal cell cultures:

1. To investigate the normal physiology or biochemistry of cells. For example, metabolic pathways can be investigated by applying radioactively labeled substrates and subsequently looking at products.
2. To test the effects of compounds on specific cell types. Such compounds may be metabolites, hormones or growth factors. Similarly, potentially toxic or mutagenic compounds may be evaluated in cell culture.
3. To produce artificial tissue by combining specific cell types in sequence. This has considerable potential for production of artificial skin for use in treatment of burns.
4. To synthesize valuable products (biologicals) from large-scale cell cultures. The biologicals encompass a broad range of cell products and include specific proteins or viruses that require animal cells for propagation. The number of such commercially valuable biologicals has increased rapidly over the last decade and has led to the present widespread interest in animal cell technology. Proteins that are present in minute quantities *in vivo* can be synthesized in gram (or kilogram) quantities by growing genetically engineered cells *in vitro*.

Advantages and Disadvantages of Using Cell Culture

In all the situations listed above we could use animal tissue (e.g. liver) or whole animals (e.g. mice). Biochemists have traditionally used homogenized liver as a source of cells for enzyme or metabolic studies. So, why use animal cell culture which may require far more time in preparation and may require specialized equipment?

The major advantage to using cell culture is the consistency and reproducibility of results that can be obtained by using a batch of cells of a single type and preferably a homogeneous population (clonal). For example, we may be doing a biochemical analysis where it is important to relate a panicular metabolic pathway to a certain cell type. This would be possible in a culture containing a homogeneous cell population which can be monitored for biochemical and genetic characteristics, but very difficult in a tissue homogenate which would contain a heterogeneous mix of cells at different stages of growth and viability.

Toxicological testing procedures have been well established in laboratory animals. However, the use of cell culture techniques may allow a greater understanding of the effects of a particular compound on a specific cell type such as liver cells (hepatocytes). Furthermore, routine toxicity tests are far less expensive in cell culture than in whole animals. In the production of biological products on a large scale, the avoidance of contaminants such as unwanted viruses or proteins is important. This can be more easily performed in a well-characterized cell culture than when dealing with pooled animal tissue.

The major disadvantage to the use of cell culture is that after a period of continuous growth, cell characteristics can change and may be quite different from those originally found in the donor animal. Cells can adapt to different nutrients. This adaptation involves changes in intracellular enzyme activities. Furthermore, culturing favours the survival of fast-growing cells which are selectively retained in a mixed cell population.

The changes in the growth and biochemical characteristics of a cell population may be a particular problem when using cultures to develop an understanding of the behaviour of cells *in vivo*. After a period of culture, the cells may be significantly different from those that are highly differentiated *in vivo* where growth has ceased. Intracellular enzyme activities change dramatically in response to nutrient depletion and by-product accumulation in a culture.

Early History of Cell Culture

The techniques necessary to allow animal cells to grow in culture were gradually developed over the course of the last century. The original impetus for the development of cell culture was to study, under a microscope, normal physiological events such as nerve development. One of the earliest experiments showing cell culture *in vitro* was conducted by Ross Harrison in 1907 who developed the

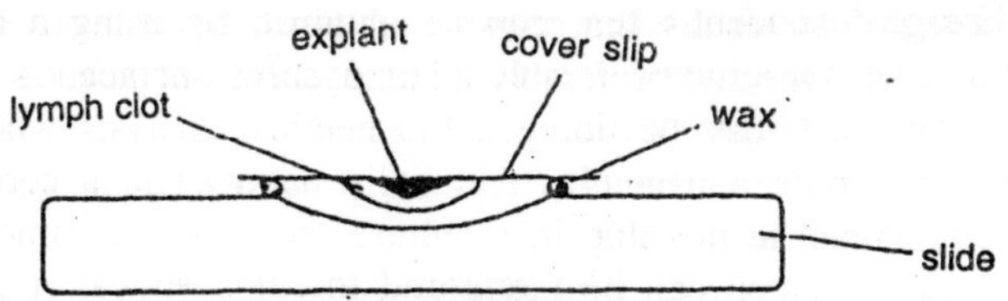

Fig. 1.1. Hanging drop technique of cell culture.

'hanging drop' technique. In the depression of a microscope slide he entrapped small pieces of frog embryo tissue in clotted lymph fluid. By this method he was able to observe the elongation of *nerve fibers* over several weeks. From his experiments he demonstrated some of the characteristics and requirements for the maintenance of cells.

Table 1.1. Milestones in the development of animal cell technology

1880	Roux maintained embryonic chick cells in saline solution
1890-1900	Harrison grew frog nerve cells by the 'hanging drop' technique
1910	Carrel used aseptic techniques for long-term cell cultures
	Rous and Jones used trypsin for subculture of adherent cells
1920	The 'Carrel' flask was designed for cell culture
1930-1940	Antibiotics were added to culture medium
	Earle isolated mouse L fibroblasts
	Enders grew polio virus on cultured human cells
1950	Gey cultured HeLa cells
	Eagle developed a chemically defined culture medium
1960	Hayflick and Moorhead showed that human cells have a finite lifespan
	Ham grew cells in a serum-free medium
	Harris and Watkins fused human and mice cells
1970	Kohler and Milstein produced an antibody-secreting hybridoma
	Sato developed serum-free media from hormones and growth factors
1980	Human insulin was produced from bacteria
	Monoclonal antibody (OKT3) used for human therapy
	Recombinant tPA licensed for human therapy
1990	Humanized chimeric antibodies used for human therapy
	Stem cells isolated

1. The cells require an anchor for support which in this case was provided by the coverslip and the fibrous matrix of the lymph clot. Cells that are bound to a tissue matrix *in vivo* are normally anchorage-dependent when grown in culture. Several proteins are responsible for glueing the cells to a solid surface. The best characterized is fibronectin.
2. The cells require nutrients which are provided by the biological fluid contained in the clot.
3. The growth rate is relatively slow (compared to bacteria) and this makes the culture vulnerable to contamination. A small number

of bacteria would soon outgrow a larger population of animal cells. The doubling time for animal cells is normally between 18 and 24 hours but bacterial cells can double in 30 minutes.

Some of these early culture experiments were continued by Alexis Carrel who was trained as a surgeon. He showed that the application of strict aseptic control enabled the prolonged subculture of the cells. He used chick embryo extracts mixed with blood plasma to support cell growth. In 1923 Carrel designed a flask suitable for routine cell culture work. This facilitated subculture under aseptic conditions and was the forerunner of the modern cell culture flask.

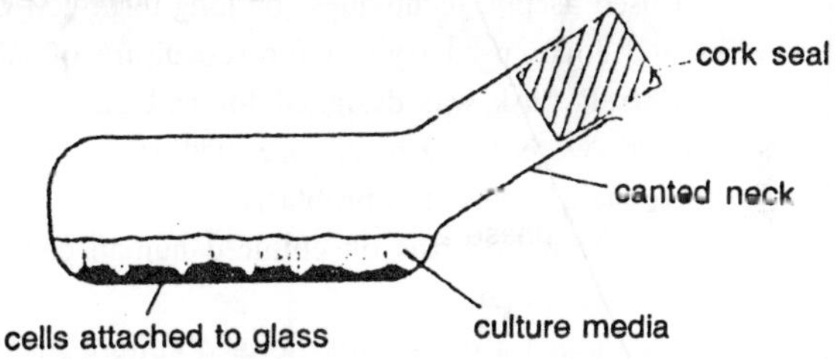

Fig. 1.2. The Carrel flask.

Sterile manipulation is facilitated by the narrow-angled glass neck of the Carrel flask which can be ilamed by a Bunsen burner prior to inoculation or sampling. However, Carrel's application of surgical procedures for aseptic manipulation of cell cultures was elaborate and difficult to repeat by others. Consequently cell culture was not adopted as a routine laboratory technique until much later. Of course, today cell culture is widely used and can be performed in most moderately equipped laboratories. Carrel is particularly well known for his 34-year culture experiment. In 1912 he isolated a population of chick embryo fibroblasts, and these cells were grown in culture until 1946. From this, Carrel concluded that the cells were immortal. However, this claim was later refuted by the work of Hayflick and Moorhead who showed the finite lifespan of isolated animal cells. It is now thought that Carrel's technique of using plasma and homogenized tissue as growth medium most likely allowed continuous reinoculation of cells into his culture.

How Long will Primary Cultures Survive?

In 1961, Haytlick and Moorhead studied the growth potential of human embryonic cells. They showed that these cells could be grown continuously through repeated subculture for about 50 generations. After

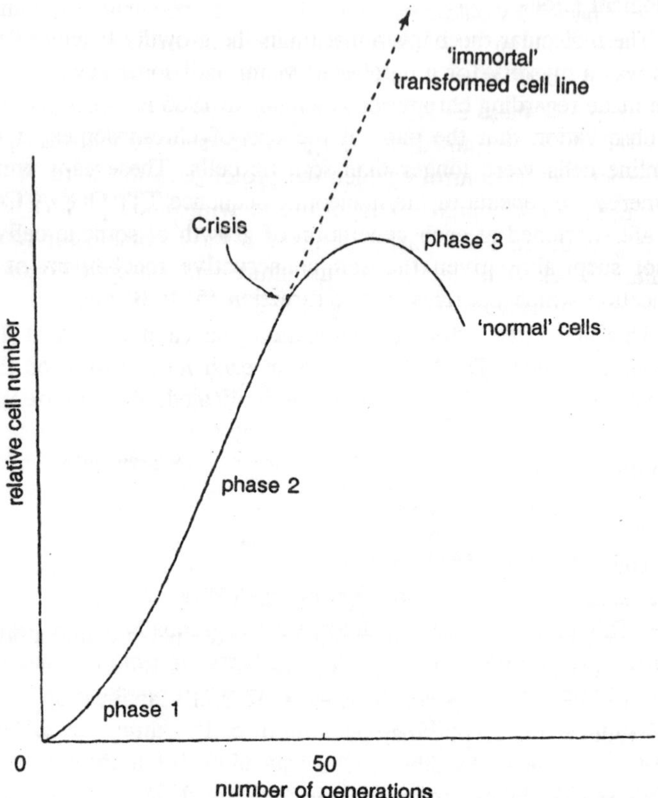

Fig. 1.3. Normal and transformed growth of human embryonic cells over an extended number of passages.

this time, the cells enter a senescent phase and are incapable of further growth. A finite growth capacity is a characteristic of all cells derived from normal animal tissue. The cells appear to pass through a series of age-related changes until the final stage when the cells are incapable of dividing further. The finite number of generations of growth is a characteristic of the cell type, age and species of origin and is referred to as the 'Hayflick limit'.

The capacity for growth is related to the origin of the cells—those derived from embryonic tissue have a greater growth capacity than those derived from adult tissue. Each cell appears to have an inherent 'biological clock' defined by the number of divisions from the original stem cell. Even if cells are stored by cryopreservation the total capacity for cell division is not altered.

Biological Clock

The molecular mechanism that limits the growth capacity of normal cells was a mystery for a number of years until some key observations were made regarding chromosomal length. In 1986 Howard Cooke made the observation that the caps at the end of chromosomes of human germline cells were longer than somatic cells. These caps known as telomeres are repeats of the nucleotide sequence TTAGGG/CCCTAA and are shortened at each generation of growth of somatic cells. This is not surprising given the semiconservative mechanism of DNA replication which operates in one direction (5' to 3' end).

This generates a fork of replication that cannot work at the end of double-stranded DNA. Therefore, at each mitotic division there is a small segment of DNA that is not replicated, thus shortening the telomeric cap. However, in germ cells the telomeree is maintained at 15 kilobases apparently by a ribonucleoprotein enzyme, telomerase. The telomerase is expressed in germ cells and has moderate activity in stem cells but is absent in somatic cells.

The finite lifespan of normal somatic cells is regulated by 10 or more 'senescence' genes that suppress the expression of the telomerase gene. This causes the human telomeres to gradually shorten at a rate of around 100 basepairs per cell division. From the telomere hypothesis, cells become senescent and stop dividing when progressive telomeric shortening produces a threshold telomere length. If oncogenes are activated then cells escape the negative control of the cell cycle and re-express telomerase. The critical experiment to validate this hypothesis came with a publication in 1998. Bodnar et al. (1998) described the introduction of human telomerase reverse transcriptase (hTRT) into normal human cells (retinal pigment epithelial cells and foreskin fibroblasts). This converted the normal $hTRT^-$ state to $hTRT^+$ with the capability for expression of the telomerase enzyme. This process succeeded in increasing the lifespan of the cells significantly.

The mean population doubling capacity of both cell types increased by at least 20 generations. For clones of the epithelial cells the mean population doubling of 54 generations increased to a value of 73 generations by the introduction of the telomerase enzyme. Similarly, for the fibroblasts, there was an increase of the mean population doubling of 64 to 100. This experiment established a causal relationship between telomere shortening and *in vitro* cell senescence which is dependent upon a mitotic clock.

First Products of Animal Cell Technology

The impetus for the application of cell culture to large-scale production came with the need for viral vaccines. Major epidemics of polio in the 1940s and 1950s prompted a lot of effort to develop an effective vaccine. In 1949, Enders showed that the poliomyelitis virus could be grown in cultures of human embryonic cells. The polio vaccine, which could be produced from the de-activated virus, became one of the first commercial products of cultured animal cells. Initially the virus was propagated on primary cells, which are defined as those taken directly from an animal into culture. Later primary monkey kidney cells were replaced by human diploid lung fibroblast cells with defined characteristics (e.g. WI-38, MRC-5). These were widely accepted because of their apparent 'normality. Since the 1950s a range of human and veterinary vaccines have been produced routinely from large-scale mammalian cell cultures.

During the 1950s, several cell types with good growth characteristics were isolated and characterized. These included the chemically transformed mouse L cells and the human carcinoma cell line, HeLa, which derives from the name of the donor - Henrietta Lach, although some texts designate the name as Helen Lane. Earle and Eagle focused their attention on the development of chemically defined nutrient formulations for the growth of these cells and to replace the undefined biological extracts previously used. The chemically defined media, Eagle's Minimum Essential Medium (EMEM) developed in the 1950s was the first widely used culture medium. EMEM had several advantages over previously used media:

1. Consistency between culture batches;
2. Ease of sterilization; and
3. Reduced chance of contamination.

Basic research in the 1960s using cell culture techniques included the classical work of Hayflick and Moorhead. This established that there is a finite lifespan intrinsic to the genetic make-up of all normal animal cells.

Recombinant DNA technology was developed in the 1970s. This allowed DNA to be isolated and inserted in any cell. Initially this was used for the expression of mammalian genes in bacteria (particularly *E. coli*). It was considered at the time that such genetically engineered bacteria could be developed for the production of any desired mammalian protein. However, it soon became apparent that large complex proteins (particularly those with carbohydrate groups attached) could only be synthesized with all their post-translational modifications in eukaryotic cells. Bacteria do not have the appropriate metabolism to complete these modifications which are necessary for full activity of many mammalian proteins. Therefore, genetically engineered animal cells were developed for the large-scale commercial synthesis of such important complex macromolecules as plasminogen activator, factor VIII and erythropoietin. These compounds have been licensed for human therapeutic use.

Fusion techniques were pioneered in the 1960s by Harris and Watkins who allowed cells of different origin to be fused to produce hybrids. The initial value of this work was to be able to assign specific genes to specific chromosomes. However, a novel application of this fusion technique was described by Kohler and Milstein in 1975 who produced hybridoma cells capable of the continuous production of a single type of antibody. These monoclonal antibodies have been particularly valued as diagnostic and therapeutic agents because of their ability to selectively bind specific compounds. They are now produced commercially in kilogram quantities from the large-scale culture of hybridomas.

The three major categories of valuable products from animal cells that we can identify over the last 40 years are:

1. Viral vaccines;
2. Monoclonal antibodies;
3. Recombinant glycoproteins.

The range of these biologicals has increased rapidly over the last few years. This is particularly so for the glycoproteins that are of use as pharmaceutical, medical or veterinary products. They are generally synthesized as secreted products from selected or genetically engineered mammalian cell lines. They are needed in large quantities and this requires a careful study of the control of the production processes. Such a study should start with an examination of the principles of biochemistry, cell biology and biochemical engineering that underlie such production processes.

2

Laboratory Equipments

The purpose of this chapter is to review some of the basic requirements for setting up a small-scale culture laboratory for handling mammalian cell lines. This will include a description of a suitable laboratory environment in which to manipulate such cells and some of the basic laboratory equipment needed.

Laboratory Design

A cell culture laboratory should allow the sterile handling of cultured cells with a minimal level of contamination. The growth rate of bacterial or fungal cells is usually so much greater than that of mammalian cells that no level of contamination can be tolerated in an animal cell culture. However, it is impractical to design a laboratory that is totally free of potentially contaminating microorganisms. The laboratory is generally designed to minimize the risk of contamination for an experienced worker. Thus, the maintenance of noncontaminated cultures is as much a product of good experimental technique as a cleat germ-free environment.

In the development of animal cell culture techniques during the 1920s and 30s, Alexis Carrel used methods borrowed from surgical practices employed in hospital operating rooms. This included full gowning with surgical dress in a sterile room. However, nowadays these methods would be regarded as rather extreme and are not normally considered necessary in a modern laboratory. To minimize the risks of contamination, several features can be incorporated into the designated clean laboratory.

An important feature of the laboratory should be a good physical separation of the sterile handling area from the areas for wash-up and sterilization. Ideally these may be separated in different laboratories. Cell culture facilities are often located in small rooms where there is minimal traffic of personnel. A small room containing a standard class-II laminar flow cabinet is ideal because the air flow through the sterile exhaust of the cabinet should maintain a low particle count in the environment. The laminar flow cabinet (hood), microscope and incubator should be positioned close together so that physical transfer of cultures will be minimized. When it is not possible to have an independent sterile handling laboratory, it is at least necessary to ensure that the sterile handling area of a larger laboratory is positioned in a region where there is minimal movement of people.

The area used for nonsterile work such as the wash-up facility and disposal of used culture flasks should be positioned at the other side of the laboratory from the dean area. This is a self-contained laboratory suitable for use by two or three peoplc. In a larger facility it would be essential to include separate rooms for the clean area and wash-up area. The incoming air into the laboratory may be filtered through a *high-efficiency particulate air* (HEPA) filter or electronic filter that may be incorporated into the ceiling space. The incoming air pressure may be increased to-cause a slight positive pressure within

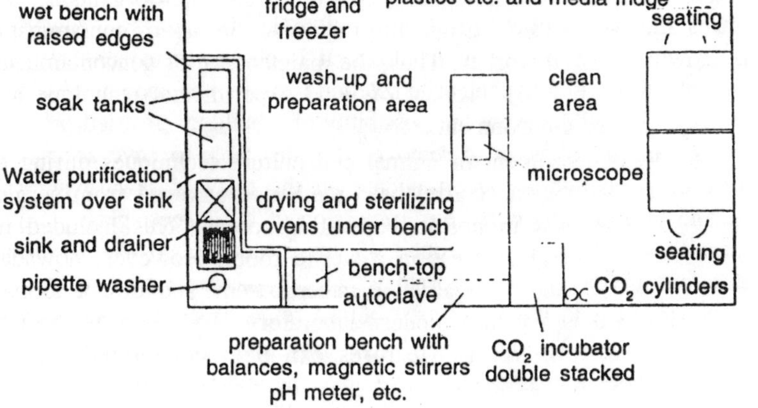

Fig. 2.1. Plan for a cell culture laboratory.

the laboratory. Air cooling ('conditioning') may often be necessary because of the heat generated by the incubators and other electronic equipment in the laboratory.

Such measures will reduce the level of potential airborne contaminants in the laboratory environment. Sterile handling of cultures is generally carried out in a laminar flow cabinet into which only the operator's arms enter the sterile work area. This reduces air currents and the potential for contamination carried by laboratory workers. This minimizes the potential of contaminants entering opened culture flasks. Other areas of the laboratory should be maintained clean by removal of unnecessary clutter and application of antiseptic cleaning agents at regular intervals.

Washing Re-usable Glassware

The general use of presterilized tissue culture-grade plasticware has meant a decrease in the amount of washing necessary in the average research laboratory involved in cell culture. It is difficult and time-consuming to wash glassware sufficiently so that it is free of potentially cytotoxic detergent to enable reproducibility in cell culture. The advantages of single-use plastic culture vessels include consistency in operation as well as a decreased preparation time. Nevertheless there is still a need for an adequate washing facility in most laboratories to prepare the spinner flasks, beakers and measuring cylinders that are used routinely. These all require a scrupulous washing regime prior to use or prior to sterilization.

Any failure in the washing operation allows the possibility of the introduction of contaminants (e.g. traces of heavy metal) into the cultures and would likely result in a cytotoxic effect causing lowered cell growth or an altered metabolism. It is necessary to soak all re-usable glassware in a hypochlorite solution (Chloros) as soon as possible after use. This will prevent the possibility of a build-up of dried protein residues that are difficult to remove. It is desirable to have access to a sink deep enough to accommodate all the dirty glassware at one time and to allow the complete immersion of the largest glass items. Generally a sink depth of around 45 cm is suitable. If the wash-up area is outside the culture laboratory then the glassware may be stored in a bucket next to the work area before removal to the main sink.

Prior to immersion in the soak tank it is advisable to rinse glassware under a cold tap, remove any tape and remove any marker pen labeling with acetone. Soaking should be for several hours or preferably overnight. In many laboratories a suitable washing machine (e.g. Lancer) is available for washing after the initial soaking. The desired procedure, either manually or by machine is to wash with a detergent and rinse with a dilute acid. Washing machine manufacturers have recommended conditions for this as appropriate for a particular machine. These procedures are followed by repeated sequential washes in tap water and then distilled or reverse-osmosis water. The glassware should be dried in a hot air oven at around 100°C before use or sterilization. Items that are not immediately sterilized should he kept covered or inverted on a clean surface until required.

Although individually packed presterilized plastic pipettes are available, many laboratories choose to re-use standard glass pipettes in cell culture operations. These may be washed and sterilized by following a similar procedure to that recommended for other glassware. Pipettes may be washed in a free-standing cylindrical pipette washer which is generally made of polypropylene. After removal of the cotton plug from each pipette, they should be placed tip uppermost into the cylinder container of the pipette washer for overnight soaking in a hypochlorite solution. This is then followed by 4-6 hours of washing using the siphon-type automatic washer. Initially this should be done with tap water and then with distilled or reverse-osmosis water. The pipettes are then dried in an oven before re-plugging with cotton wool and sorting pipettes of the same size into metal cans. These may be sterilized by dry heat (160°C for 1 hour).

Biosafety/Laminar Flow Cabinets

Sterile operations in a cell culture laboratory are normally undertaken in cabinets (commonly called '*hoods*'), which serve the purpose of minimizing the chance of culture contamination and ensure the safety of the operator. For media preparation, or when handling nonprimate cell lines, culture manipulations can be conducted in a small front-opening cabinet. The cabinets are relatively cheap and are equipped with an ultraviolet (UV) light source to prevent contamination of the cabinet surface when not in use but may not have an internal air flow. The surface of the cabinet should be disinfected regularly and bottles or flasks should not be stored in the cabinet. Good technique will ensure that cultures do not become contaminated. However, it must be noted that this system does not offer operator protection against possible pathogens.

If human or other primate cells are used then some protection is required against the possibility of transmission of infectious agents. For this purpose most cell culture laboratories have an open-fronted laminar flow cabinet in which only the operator's arms enter the sterile area. The cabinet offers a space containing a vertical flow of filtered air and a horizontal working surface that can be disinfected. The most commonly used cabinet for cell culture is 1.22 m long and designated as class II. The classification is a measure of the biological safety.

A class II cabinet is suitable for work with low-to-moderate toxic or infectious agents. There is an inward flow of air drawn into the sterile working area of the cabinet through a high-efficiency particulate air (HEPA) filter. The direction of flow offers personnel protection. Most of the air (70-80%) is recirculated to form an air curtain which serves to maintain sterile space for culture manipulation. The air curtain funned in the internal front face of the cabinet has a typical flow rate of 0.4 m/s. The exhausted air is also forced through a HEPA filter and this serves to protect the surrounding environment from any potential pathogens or toxic compounds. The design of the class II cabinet is such that there is free access for the operator's hands but a Perspex cover prevents the operator breathing over the working surface.

The HEPA filters are designed to trap extraneous airborne particles or aerosols. They are constructed from a continuous sheet of submicron glass fiber folded back and forth over a corrugated spacer as a support. The common HEPA filter in a class II cabinet ensures a 99.99%

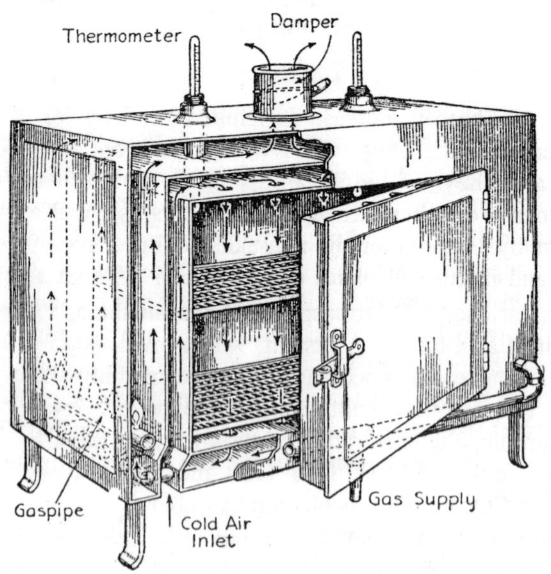

Fig. 2.2. Hot-air sterilizer.

efficiency of entrapment of 0.3 μm particles. The cabinet should be located in a corner of the laboratory free from draughts and air movement. A source of UV light is an optional feature that can be built into these cabinets. The purpose of the UV light is to maintain sterility when the cabinet is not in use. It is also advisable to spray and wipe the horizontal working surface of the cabinet with 70% ethanol before use. The class III cabinet is a totally enclosed system found in specialized laboratories designed to handle high-risk pathogens. This type of cabinet is completely sealed and contains glove pockets to allow manipulation of the cultures. The exhausted air from the cabinet is filtered through at least two HEPA filters to ensure the complete removal of all pathogens. All equipment entering the cabinet is passed through an air lock and removed directly into an autoclave. The class III cabinets are required for handling highly pathogenic material such as virus-producing human cell lines or tissue samples carrying known human pathogens.

INCUBATORS

Incubators are a basic necessity for maintaining a constant temperature during cell culture, usually at 37°C for mammalian cells. These incubators may range from temperature-regulated boxes to more elaborate systems that can control temperature, humidity and CO_2. These three parameters are essential for maintaining the consistency of culture conditions as well as preserving the viability of the cells. A slight decrease in temperature from the optimal may slow the cell growth rate but an increase in temperature is likely to be far more detrimental to the cells. Cells will survive at 39°C only for a few hours and they will die rapidly at a temperature above 40°C. Maintenance of a humidified atmosphere is essential to prevent loss of medium by evaporation in nonsealed culture systems such as Petri dishes and multi-well plates or in T-flasks without an airtight cap.

An enriched CO_2 atmosphere of the incubator chamber is usually maintained as a buffering system, which involves an equilibration with the bicarbonate contained in the culture medium. The required level of CO_2 to maintain a culture pH at around 6.9-7.4 will depend upon the bicarbonate concentration of the medium. Most growth media are rated for either 5% or 10% CO_2 and this necessitates a constant supply of CO_2 into the incubation chamber.

Abnormal CO_2 levels in the incubator are reflected by a change in pH of the culture medium as observed by the colour. Too low a level of CO_2 causes an increase in culture pH whereas too high a level of CO_2 causes a decrease in pH. The CO_2 level can be monitored by a Fyrite test kit. In this simple apparatus a gas sample is pumped manually into an absorbing liquid by a rubber bulb. The level the liquid rises in the glass column indicates the % CO_2.

Table 2.1. Recommended levels of CO_2 in incubators with different media

Bicarbonate in medium (mM)	*Recommended CO_2 level in gas phase*	*Example of medium*
4	Atmospheric	Hank's BSS
26	5%	RPMI 1640
44	10%	DMEM

The disadvantage of the bicarbonate-CO_2 buffer system is that cultures may become alkaline very quickly when removed from the incubator. It is also possible to use an organic buffer to maintain culture pH. HEPES (N-2, hydroxyethylpiperazine N-2-ethancsulfonic acid; pK_a –7.3 at 37°C) or MOPS (morpholinopropane sulfonic acid; pK_a = 7.0 at 37°C) at a concentration of 10-20 mM will maintain culture pH without an enriched CO_2 atmosphere. In the presence of HEPES the CO_2 level can be reduced to around 2% with a concomitant decrease in bicarbonate concentration. The organic buffers may be used in addition to the bicarbonate-CO_2 buffering system when a high degree of pH control is required. The disadvantage of using organic buffers is that the medium becomes expensive and therefore they are not generally used for routine culture.

The CO_2 incubator in which a fixed CO_2 tension is maintained in a humidified atmosphere has now become a standard piece of equipment, in a cell culture laboratory. In a small laboratory the CO_2 cylinders are secured to a rack alongside the incubator. Gas is fed via a reducing valve on the cylinder head through pressure tubing to an intake port that passes through a filter before entering the incubation chamber. For older designed CO_2 incubators there is a constant flow of a CO_2 /air mixture with each gas supplied separately from a different cylinder and mixed to set proportions via gas burettes. This has the disadvantage of excessive usage of CO_2 and if the CO_2 supply runs out then the constant air flow flushes out any remaining CO_2 from the incubator.

In a modern system two CO_2 cylinders are mounted alongside,each incubator. Each cylinder has two regulator valves. One is a high-pressure gauge that measures the pressure in the cylinder with a range

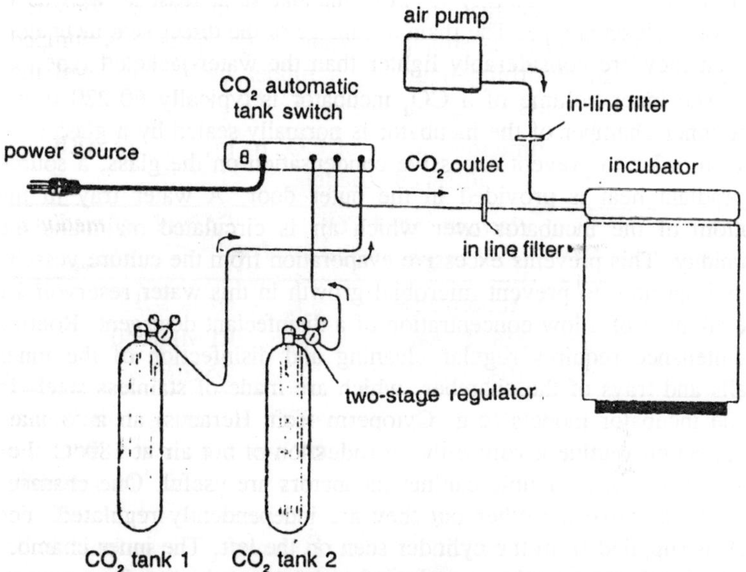

Fig. 2.3. A typical gas control system for a modern CO_2 incubator.

of 0-2000 psi and a second is a low-pressure gauge (0-30 psi) measuring the gas pressure into the incubator. Pressure tubing connects the low-pressure gauges of the two cylinders to a switching device, which if automated will switch the CO_2 supply from cylinder one to cylinder two if the pressure falls below a certain level (typically 4 psi). This automated switchover unit is essential to ensure a constant CO_2 supply.

The % CO_2 in the incubation chamber is controlled by a valve governed by intermittent readings of an infrared gas analyzer. Such a sensor system is built into the control box of the incubator and is capable of maintaining a constant CO_2 level to an accuracy of ±0.1%. The infrared controller measures the CO_2 level independently of humidity and incorporates a correction for temperature.

Air is normally pumped into the incubation chamber by a small pump via a filter. The circulation of air ensures an even temperature throughout the chamber. The chamber temperature is effectively controlled by a large jacket through which the circulation of water can ensure a uniform temperature of ±0.2°C. Other temperature control devices can be used but traditionally they have been assumed to be not as good as water jackets in ensuring an even chamber temperature. However, modern direct heat CO_2 incubators are available with heating elements on all sides of the outer chamber wall. Microprocessor control can ensure thermal stability in these incubators at least as well as in the water-jacketed type. The main advantage of the direct heat incubators is that they are considerably lighter than the water-jacketed type.

The inner volume of a CO_2 incubator is typically 60-220 liters. The inner chamber of the incubator is normally sealed by a glass door and in order to prevent excessive condensation on the glass, a source of radiant heat is provided in the outer door. A water tray in the bottom of the incubator over which air is circulated maintains the humidity. This prevents excessive evaporation from the culture vessels. It is important to prevent microbial growth in this water reservoir by the addition of a low concentration of a disinfectant detergent. Routine maintenance requires regular cleaning and disinfection of the inner walls and trays of the chamber, which are made of stainless steel. In some incubator models (e.g. Cytoperm from Heraeus) an automated disinfection routine involves the introduction of hot air at 180°C. For many laboratories double cabinet incubators are useful. One chamber is stacked above the other but they are independently regulated. The CO_2 is supplied from the cylinder seen on the left. The inner chamber of the incubator contains a series of racks onto which culture vessels can be placed.

Laboratory-scale Culture Flasks

In the past, culture flasks were made of borosilicate glass. The most common was the Roux bottle consisting of two flat sides with a typical surface area of 230 cm^2 per side for the attachment and growth of anchorage-dependent cells and a capacity for a 500 ml culture volume. The flasks were re-cycled and required washing and autoclaving before use—a process that could give rise to contamination. However, presterilized plastic flasks suitable for cell culture are now commercially available and are used by most laboratories. The plastic is polystyrene, which is treated to produce a surface amenable for cell attachment and growth. The tissue-grade plastic flasks are sterilized by gamma irradiation and are suitable for single use. The use of presterilized and disposable plastic flasks has significantly reduced any contamination arising from the culture vessels. There is no need for an extensive washing process, which is critical to ensure the complete removal of cytotoxic contaminants from glass containers.

Whatever material is used, the surface charge density is critical for the attachment of cells. Thus, the physical and chemical treatment of surfaces has a considerable effect on cell adhesion. New borosilicate glass flasks or bottles often show poor cell adherence but this can be improved by washing, sterilization and chemical treatment. An appropriate charge can be placed on the glass by alkali treatment. Typically, this involves addition of 0.1 M EDTA in 25 mM NaOH at 122°C for 30 minutes followed by washing with sodium carbonate. The negative charge on the surface of glass can be manipulated by the extent of alkali treatment. The effect of alkali treatment is to rupture the Si-O-Si bonds of the silica network forming SiO^- residues with counter ions of Na^+ on the surface structure. The depth of penetration of Na^+ depends on the extent of treatment. The state of the surface can be quantitatively assessed by the extent of adsorption of the positively charged quaternary ammonium dye, crystal violet. Cell attachment can be improved by increasing the negative charge to a level specific for each cell type.

The attachment of the negatively charged cell surface to a negatively charged substratum requires the presence of a divalent cation such as Ca^{2+} and Mg^{2+} in the culture medium. Electron microscopic analysis of the cellsurface interface shows the presence of a 50 Å protein layer between the cells and the substratum. This may originate from surface-active proteins secreted by the cells or from a serum supplement in the growth medium.

The polystyrene used to make most domestic plastic containers is unsuitable as a surface for cell attachment because it is hydrophobic and has no charge. The 'tissue culture grade' plasticware available from commercial sources is generally made of polystyrene, which is treated to allow a negative surface charge. This provides a surface chemistry that is hydrophilic, wetable and negatively charged. The charge may be provided by short exposure of the polystyrene to sulfuric acid to allow sulfonation of the surface. Alternatively, the usual commercial process is corona-oxidation in which the plastic is exposed to a high-voltage electric arc. This also leads to a surface with a layer of negatively charged groups. There has been shown to be an optimal surface charge for cell attachment at a negative charge density of $2\text{-}10 \times 10^{14}$ charges per cm^2. The polystyrene growth surface may be further modified by applying a variety of anionic and cationic groups. The polymeric surface with both amine and carboxyl groups is available in sterilized culture flasks as Primaria products (Falcon). These surfaces are optimal for the growth of certain specialized cell lines.

Cells have also shown an ability to attach to various positively charged polymers. These include DEAE-dextran, polyacrylamide, polylysine, polyornithine, polyarginine, polyhistidine and protamine. Thus, although the charge density on the surface appears to be a critical parameter, the charge may be positive or negative. Poly-D-lysine of molecular weight in the range 30-300 kDa or 30,000-300,000 Da has been used extensively as coating polymer for certain applications to provide a positive charge on glass or plastic. Washing a flask with a polylysine solution (10 μg/ml) is generally sufficient to apply the polymer coating. This may be useful to encourage the attachment of cells that usually grow in suspension. The surface charge density can also be manipulated by choosing a selected molecular weight of the polymer.

The choice of culture vessel for the growth of cells is dependent on the scale of operation and whether suspension or anchorage-dependent cells are required. Culture volumes from 10 μl to 10,000 liters are possible with existing commercially designed equipment. The smaller volume vessels are generally not equipped with control devices and it is accepted that pH and oxygen concentrations in culture may fluctuate during cell growth. However, the advantages of these vessels are that they can be handled in replicates and they are suitable for insertion into CO_2-enriched incubators. Most of these vessels offer a flat surface for cell attachment, although cells may be grown in suspension or surface-attached.

The most popular forms of plastic culture containers are multi-well plates, Petri dishes and flasks (usually referred to as Tissue culture flasks or T-flasks) made of tissue culture-grade polystyrene. The multi-well plates can accommodate many replicates of small-volume cultures. The 24-well plates hold 3 ml per well and are well suited for cell growth experiments, for example to test for toxicity or stimulatory activity. The 96-well plates hold a volume of 0.3 ml per well and are suitable for cloning or replicate assays. Rapid dispensing of solutions into the 96-well plate is made easy by use of a multi-well pipettor. These are lightweight and normally have an adjustable volume setting. The pipettors are designed for use with sterile pipette tips. The 96-well plate reader can be used to obtain simultaneous optical density data from all the wells from a single plate.

For some applications porous inserts for multi-well plates are available. These have polycarbonate or polyethylene terephthalate membranes of various pore sizes (0.4-8 μm). Cells can grow over the membrane provided. This may be useful to study cell polarity as access is provided to the basolateral and apical sides of the cell layer.

The Petri dishes and T-flasks can accommodate cultures of 2-100 ml and are suitable for both anchorage-dependent and suspension cells.

The T-flasks are designated by the surface area available for cell attachment. Thus T-25, T-75, T-150 and T-175 flasks have a growth area of 25-175 cm^2. A canted (angled) neck is provided so that sterile manipulation is easy. It is important to allow an equilibrium to develop between the gas phase of the flask and the atmosphere of the incubator chamber. The T-flask caps can be adjusted to fit loosely, to allow gas exchange with the environment of the incubator. The caps should be closed tightly once the flasks are removed from the incubator. Some T-flasks have a ring of five holes in the plastic caps to allow gas exchange through an inner permeable membrane.

Spinner bottles are straight-sided glass flasks containing a suspended central Teflon paddle containing a magnet, which turns and agitates the culture when placed on a magnetic stirrer. The stirring should be stable over a long period at a rotation speed of between 10 and 300 rpm. The bottle (or flask) is usually fitted with one or more side arms. These are useful for sampling or as ports for probes or tubing. Spinner bottles can be designed up to a capacity of 5-10 liters. Cultures above this volume require a top-driven motor for stirring.

The spinner bottles are suited for growing suspension cells although they can be adapted for anchorage-dependent cells by the use of microcarriers. The spinner flasks are usually siliconized to prevent

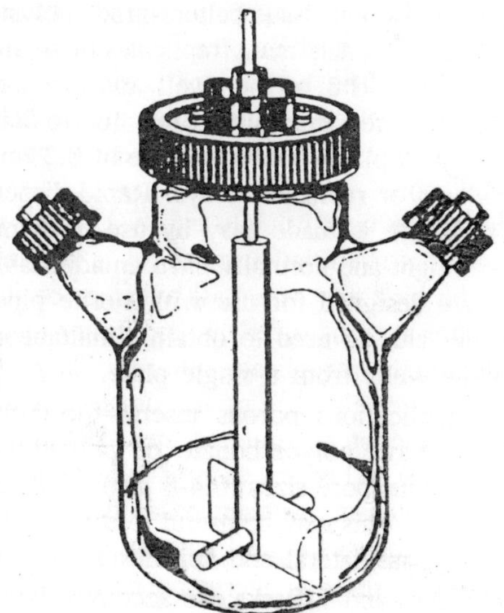

Fig. 2.4. Spinner bottle with suspended paddle.

undue attachment of cells to the inner glass surface. This may be performed by application of dimethyldichlorosilane. An appropriately designed stirring base is required for operation of the spinner bottles. Most laboratory stirring bases used for chemical operations are unsuitable because they have poor speed control and cause excessive heating. The stirring bases used for cultures are required to maintain good speed control for long time periods and without excessive heating. The stirring bases are normally, fitted with a tachometer for an accurate indication of the stirring rate. Although these cultures are often established in warm rooms many CO_2 incubators will also accommodate such a stirring base—the power line may be fitted through a sealable hole provided at the side of the incubator. However, prolonged use of stirring bases in a humidified incubator can cause corrosion of the internal mechanical drives. This problem has been solved by some advanced stirrer modules, which are sealed units placed inside the humidified incubator. A separate master controller that can control several stirrer modules is placed outside the incubator. A larger surface area for cell growth is offered in roller bottles, which are cylindrical plastic containers that are placed on their side onto mechanical rollers. The three standard sizes of roller bottles offer a growth surface of 490, 850 or 1750 cm^2 and can be filled with 250-500 ml of medium.

The rolling mechanism allows the bottles to be rotated gently and the culture medium to flow continuously over the inner surface. On inoculation into the culture medium, the cells attach and grow over the entire inner surface. The bottles are positioned to turn slowly along the long axis at between 5-60 revolutions per hour (rph). The volume of medium added should be just sufficient to provide a shallow covering of the cell monolayer. After each complete turn of a bottle the entire cell monolayer is transiently exposed to the medium. It is important to ensure that the rotation platform is perfectly horizontal otherwise cells at the high end of a flask will not be in contact with medium. Roller bottle systems offer the possibility of high yields of anchorage-dependent cells in replicate cultures. They were originally developed for the large-scale culture of anchorage-dependent cells used in the commercial production of viral vaccines. Equipment is available to accommodate up to 30,000 bottles each of 1 liter capacity. However, the process is labour intensive as each roller bottle must be handled individually for media changes or cell harvest.

Microcarrier cultures offer an alternative means of producing large quantities of anchorage-dependent cells and these have now been adopted in most commercial processes for vaccine production. An alternative culture vessel for producing a large yield of cells on a laboratory scale is the multi-tray unit called the *Cell Factory* (Nunc). This is a plastic container that incorporates 10 large plastic trays sealed together. The total growth surface area is 6000 cm^2. No rolling is required and a maximum yield of over 109 anchorage-dependent cells can be expected. The system is designed for single use.

Microscopes

An inverted microscope is essential for examination of a cell culture at regular intervals to monitor the health and growth of cells. The design of this microscope with the light source at the top and a long working distance condenser allows cells in flasks or even roller bottles to be viewed. Changes in the cell morphology, granularity and degree of spreading are all indicators that can be monitored by regular microscopic examination of cultures. A standard microscope with a movable slide holder is required to count cells in a culture sample introduced into the counting chamber. In addition to this, it is worthwhile inspecting culture samples under a microscope to determine if there are any significant changes to the appearance of the cells. Both types of microscope are essential for routine monitoring of cells in culture.

Centrifugation

A low-speed centrifuge is a requirement to harvest cells from culture. In general a centrifugal force of 150-200 g for 5-10 minutes should be sufficient to separate out cells from culture medium. Higher forces or longer times may cause damage to the cells by compacting them at the bottom of the centrifuge tube. Once centrifugation is completed, the supernatant should be decanted and the cells re-suspended. Typically swing-out buckets in a bench centrifuge will accommodate 50 ml or 15 ml plastic centrifuge tubes. The bucket size required will depend upon the volume of culture to be handled.

For reasons of safety the centrifuge rotor or individual buckets should be scalable to contain any spillage or breakage. Also, the centrifuge chamber should be sealed. Fine control of braking is desirable particularly if isolating cells from a concentration gradient. Gentle braking prevents disruption of the separated bands. A benchtop microfuge may also be useful for higher-speed centrifugation of small volumes of reagents or media samples. These may develop precipitates after freezing. Also, many analytical techniques performed on cell culture media require precipitation of proteins prior to analysis.

Liquid Nitrogen Storage

Cells can be stored for long periods at sub-zero temperatures. This permits cell stocks to be maintained in laboratories without having to resort to primary animal tissue. The maintenance of a cell stock guards against loss of a cell line by contamination or by genetic change. For cells that are grown continuously it is often desirable to store cells after various passages so that any genetic change can be monitored.

For a valuable cell line it is common to maintain a two-tiered cell bank—a master cell bank and a working cell bank. The master bank is a store of cells at early passage and established soon after receiving the original cells. The working bank is a store of cells formed by growth for several passages of one of the master bank samples. Future cell samples for inoculation of cultures are taken from the working cell bank and the master cell bank is accessed only when absolutely necessary. This practice is common in industrial production when the lifespan of the cell lines is carefully monitored.

Cells can be stored in a suspension (10^7 cells/ml) in a freezing medium that is dispensed into plastic ampoules (typically 2 ml). The freezing medium consists of growth medium or serum supplemented with a cryoprotectant such as 10% glycerol or dimethyl sulfoxide (DMSO) that will protect the cells from disruption during the freezing and thawing process. The cells are stable almost indefinitely in the cryo-protectant when held under liquid nitrogen (–196°C). Most cell culture laboratories will have liquid nitrogen storage canisters for such a purpose. The method of freezing and thawing is important to maintain a high viability of the stored cells. Slow freezing and rapid thawing is recommended for maximum cell survival.

The cell suspension can be frozen by placing the ampoules containing the cells in a polystyrene box held at –70°C overnight. This ensures an initial freezing rate of about 1°C/min after which time the ampoules are placed directly into liquid nitrogen. Programable coolers are available to control the rate of cooling. These are based on a slow infusion of liquid nitrogen at a rate determined by the pre-set cooling program. However, they are not widely used in research laboratories because of the expense and the fact that there is no great advantage unless it is required to vary the rate of cell freezing. For recovering cells, the ampoules are transferred as quickly as possible from liquid nitrogen to a water bath at 37°C. The water bath should be covered for safety because there is a possibility that a damaged ampoule may explode if liquid nitrogen has penetrated the seal.

A liquid nitrogen freezer can range in capacity from 25-500 liters and may be narrow necked or wide necked. The freezers require regular addition of liquid nitrogen to maintain their temperature. The wide neck type of freezer has the advantage of easy access but the rate of liquid nitrogen evaporation is higher. The cells are stored in plastic vials (1-2 ml) which are lowered into liquid nitrogen for long-term storage. These are placed in large drawers (for wide-necked freezers) or attached to metal canes (for narrownecked freezers). The cryogenic vials are contained in the plastic boxes, which are stacked in the metal racks before lowering into the liquid nitrogen. Depending on the size of the freezer, the storage capacity will vary between 250-15,000 plastic vials (or ampoules). A capacity for 1200-1500 vials is more than adequate for most research laboratories. Cryogenic plastic vials have strong seals to prevent leakage that could result from large temperature fluctuations. The liquid nitrogen reservoir needs to be replenished at regular intervals. This requires consistent and regular monitoring of the liquid nitrogen content in the reservoir, so that the stored vials of cells are maintained frozen. This may be performed with a dipstick. However, many modern freezers are fitted with an automatic indicator with an alarm, which sounds if the liquid nitrogen level gets too low. This lowers the risk of damaging important cell stocks by inadvertently allowing the liquid nitrogen level to drop too low.

Osmometer

An important parameter of cell culture medium is the osmotic pressure expressed as osmolarity (number of particles per liter) or osmolality (number of particles per kilogram). Most measuring devices will determine the osmolality which approximates to the osmolarity in dilute solution. The effect of each component in the medium to the overall osmolarity is additive and dependent upon its dissociation. Therefore, 15 mM glucose will increase the osmolarity by 15 mOsm/l whereas 15 mM NaCl will increase the osmolarity by 30 mOsm/l.

The osmolarity of standard culture medium is approximately 300 mOsm/l and is optimal for most cell lines. Cells can normally tolerate variations within 10% of this value. However, care should be taken in adding supplements to the media as the osmolarity may be adversely affected. The osmolarity of a culture may increase during cell growth as a result of the production of low-molecular-weight metabolites such as ammonia and lactic acid. An off-line measurement of the osmolarity of the culture may be made with a simple bench-top osmometer. The most common type is based on the measurement of the freezing point of the liquid. The principle is that the freezing point is lowered as the total number of all dissolved particles (ionic and nonionic) is increased. Water has a freezing point of 0°C whereas a saline solution with an osmolality of 1 Osmol/ kg has a freezing point of -1.858°C. Osmometers will measure the freezing point of a solution in comparison to that of water to an accuracy of ±0.001°C. Most models will measure within a range of 0-3000 mOsm/kg. Variations between different models of osmometers include the degree of automation for measuring multiple samples and the sample size required. In a micro-osmometer a sample size of 20-50 μl would be sufficient whereas sample sizes for a standard instrument are 0.2-2 ml.

3

CHARACTERISTICS OF CELLS IN CULTURE

A major choice has to be made when establishing a cell culture as to whether cells are obtained directly from animal tissue or from a culture collection. The choice will depend upon what the objective of the project is and the nature of the experiments planned. Isolation directly from tissue offers a means of culturing cells close to their state *in vivo*. However, the isolation process is far more demanding and troublesome compared to establishing a culture from a cell sample that could be obtained from a culture collection. Most culture laboratories prefer to use cell lines from collections because they are well characterized in terms of growth, origin and genetic traits.

Cells from Tissue: a Primary Culture

A primary culture is established when the cells taken directly from animal tissue are added to growth medium. Primary cultures are often established from embryonic tissue because the cells are more easily dispersed and have a superior growth potential. The structure of a tissue is highly ordered and comprises a range of cell types as shown for the cross-section of the epidermis. The objective of establishing a primary culture is to select a single cell type from this ordered structure.

The original methods developed for tissue culture by Harrison and Carrel involved the maintenance of tissue fragments (or explant) on a solid surface and supplied with suitable nutrients. However, such cultures are of greater use if the individual cells are separated out before culture. This is done by fragmenting tissue with forceps and

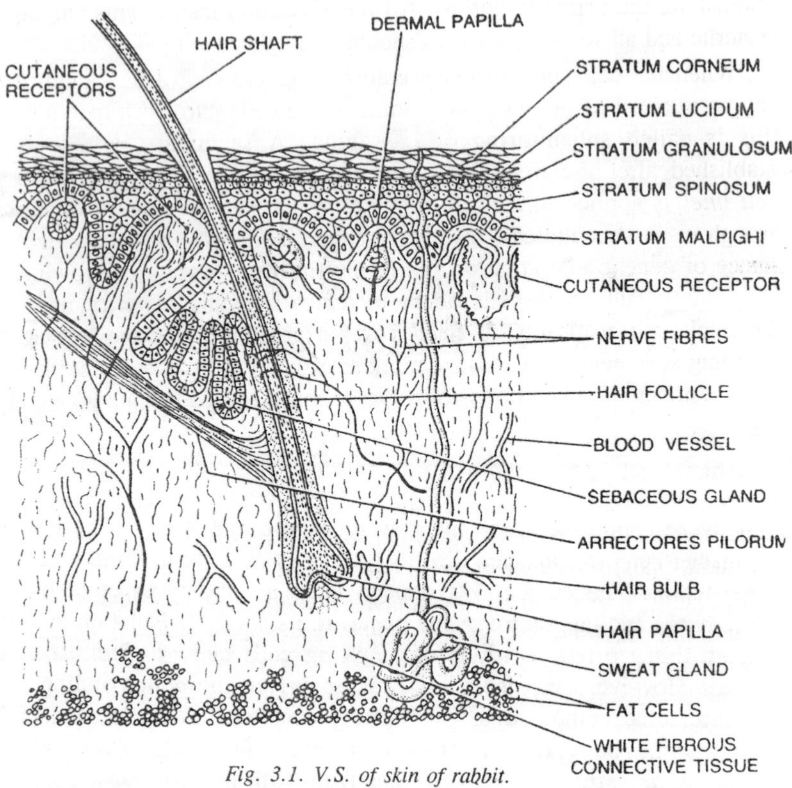

Fig. 3.1. V.S. of skin of rabbit.

scissors followed by treatment with a proteolytic enzyme such as trypsin or collagenase. The proteolytic action of the enzymes disaggregates the tissue into individual cells after which the cells are isolated by low-speed centrifugation. The time that the cells are in contact with the degradative enzymes should be minimized otherwise membrane damage may occur. Cells can be bathed in trypsin for 10-20 minutes. Longer exposure times are undesirable because breakdown of the protein components of cell membranes could occur. Collagenase degrades collagen and is less harmful to cell membranes.

This technique works well for most tissues although some modifications to the general procedure may be necessary to ensure the maximum yield of a particular cell type. One of the major difficulties and reasons for failure at this stage is that the cell population becomes contaminated with bacteria or fungi. To avoid this problem it is important to maintain aseptic techniques throughout the process of establishing the primary culture. All the dissection instruments should be sterile and all working surfaces should be swabbed with 70% alcohol.

When the cells in a primary culture stop growing a new culture may be established by inoculating some of the cells into fresh medium. This is called subculturing or passaging. A secondary culture is established after the first passage of the primary culture. The term '*cell line*' is applied to the cell population that can continue growing through many subcultures. However it should be noted that the greatest chance of genetic alteration occurs in the first few passages following the primary culture as cells adapt to a new chemical environment. The chick embryo fibroblasts may grow for around 30 passages before becoming senescent. The passage number of a culture is often recorded as the number of subcultures from the primary source.

Cell Types

Animal cells are usually defined by the tissue from which they have derived and have characteristic shapes that can be observed and recognized easily through a light microscope. These are derived from five main types of animal tissue.

1. Epithelial tissue consists of a layer of cells which cover organs and line cavities; examples include skin and the lining of the alimentary canal. The epithelial cells grow well in culture as a single cell monolayer and have a characteristic cobble-stone appearance.

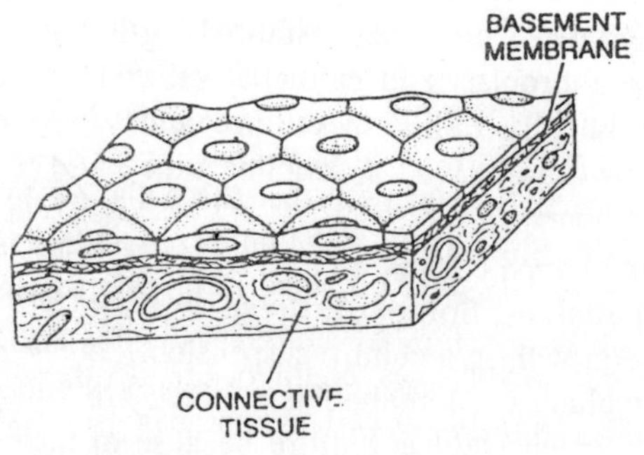

Fig. 3.2. Simple squamous epithelium.

2. Connective tissue forms a major structural component of animals, consisting of a fibrous matrix and including bone cartilage. The tissue contains fibroblasts which are amongst the most widely used cells in laboratory cultures. Fibroblasts are bound to the fibrous protein collagen in the connective tissue. The cells are spherical

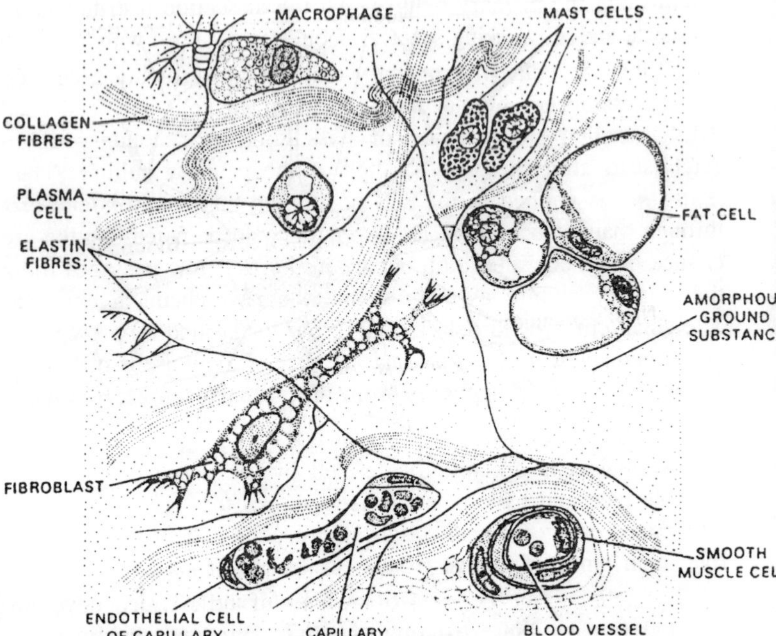

Fig. 3.3. Areolar connective tissue.

when first dissociated by trypsin from the tissue but elongate to a characteristic spindle-shape on attachment to a solid surface. Fibroblasts have excellent growth characteristics and have been the 'favourite' cells for establishing cultures. Fibroblast and epithelial cells adapt relatively easily to culture and have growth rates with a doubling time of 18-24 hours.

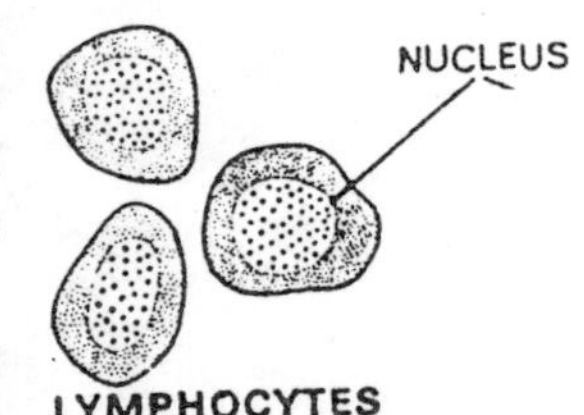

Fig. 3.4. Lymphocytes.

3. Blood and lymph contain a range of cells in suspension. Some of these will continue growth in a culture suspension. These include the lymphoblasts which are white blood cells and are used extensively in culture because of their ability to secrete immunoregulating compounds.
4. Nervous tissue consists of characteristically shaped neurons which are responsible for the transmission of electrical impulses and supporting cells, such as glial cells. Neurons are highly differentiated and have not been observed to divide in culture. However, the addition of nerve growth factor to cultures of neurons

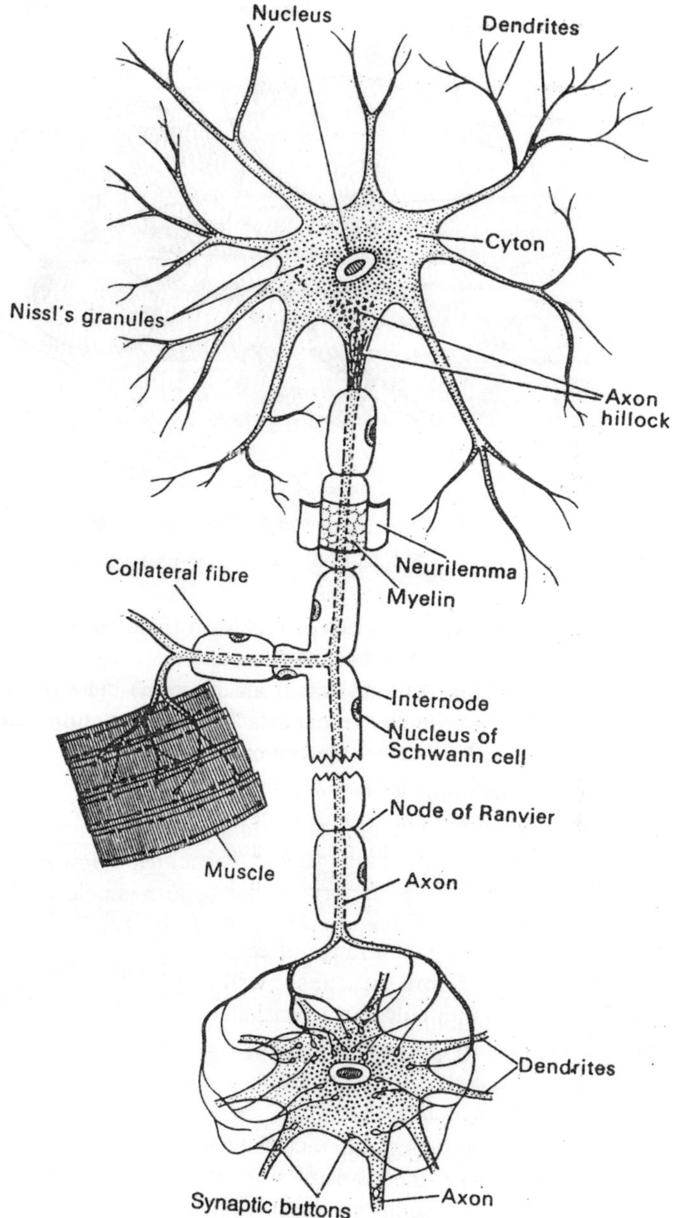

Fig. 3.5. An enlarged neuron and its synapses with another neuron.

may cause the formation of cytoplasmic outgrowths called neurites. Some of the characteristics of nerve cells can be observed with

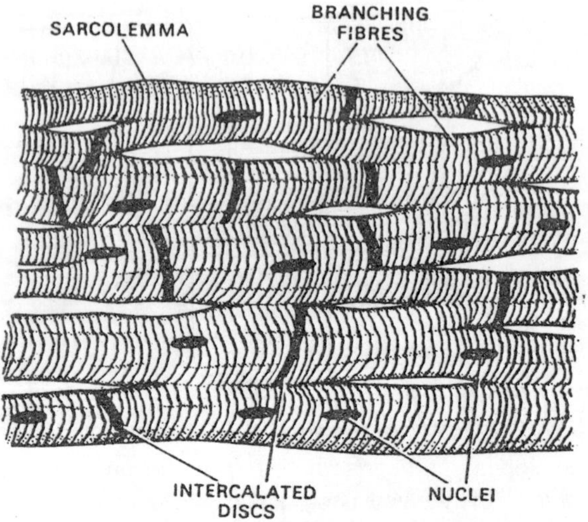

Fig. 3.6. The "syncytial" interconnecting cardiac muscle.

neuroblastomas which are tumour cells that undergo cell growth in culture.

5. Muscle tissue consists of a series of tubules formed from precursor cells which fuse to form a multinucleate complex and which also contain the structural proteins (actin and myosin). The precursor cells: are myoblasts which are capable of differentiation to form myotubes—a process that can be observed in culture

How to Select a Particular Cell Type

The primary culture will almost certainly contain a variety of different cell types with differing growth capacities. However, for most experimental work it is important to isolate a single cell type from the culture population. There are several ways this can be achieved.

1. Allow the cells to grow. Fast-growing cell types may assume dominance in a population. For example, fibroblasts have relatively short population doubling times and may outgrow other cells after a few generations .(called 'fibroblast overgrowth').
2. Control the composition of the growth medium. The addition of specific growth factors of known growth inhibitors may allow selective growth of certain cell types.
3. Separate cells by using gradient centrifugation. The cells sediment to an equilibrium position equivalent to their own density—a process called *isopycnic* sedimentation. The gradients can be formed by nontoxic, high molecular-weight material such as colloidal silica

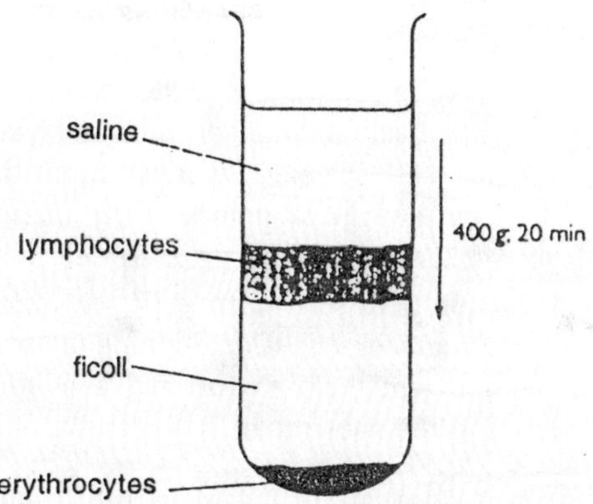

Fig. 3.7. Separation of cell by density gradient centrifugation.

as developed by Pharmacia in their formulations '*Ficoll*' and '*Percoll*'. This method is particularly effective for the isolation of certain cell types in sterile medium, for example lymphocytes from blood.

What is a 'Normal' Cell ?

In the 1960s, 'normal' mammalian cells were required as hosts for the production of human vaccines in order to ensure the safety of these products. In order to meet this requirement, a number of characteristics of 'normal' animal cells were defined by Hayflick and Moorhead following their work with human embryonic cells:

1. A diploid chromosome number (e.g.: 46 chromosomes for human cells). This indicates that no gross chromosomal damage has occurred;
2. *Anchorage dependence*. The cells require a solid substratum for attachment and growth. Growth continues until a confluent monolayer of cells is formed on the substratum;
3. *A finite lifespan*. This is a reflection of the intrinsic growth potential of the cells;
4. *Nonmalignant*. The cells are not cancerous. This can be shown by the inability of the cells to form a tumour following injection into immunocompromised mice.

Anchorage-dependence

Anchorage-dependence is the requirement of cells for a solid substratum for attachment before growth can occur. At the laboratory scale this substratum can be provided by the solid surface of Petri dishes, T flasks, or Roux bottles which are made of specially treated glass or plastic. The interaction between the cell membrane and the growth surface is critical and involves a combination of electrostatic attraction and van der Waal's forces. Cell adhesion occurs by divalent cations (usually Ca^{2+}) and basic proteins forming a layer between the solid substratum and the cell surface. In most cases the cellsurface interaction is provided by a range of nonspecific proteins which form a 2.5 nm-thick layer on the substratum prior to cell attachment.

Serum-derived glycoproteins (e.g. *fibronectin*) can provide a surface coating conducive to cell attachment. Conditioning factors are released by cells into the medium and help in forming a bond between cell surface glycoproteins and the substratum.

The density of the electrostatic charge on the solid substratum is also critical in maximizing cell attachment. A negative charge is provided on glass surface containers by alkali treatment. Tissue culture-grade plasticware consists of sulfonated polystyrene with a surface charge of 2-5 negatively charged groups per nm^2.

Culture systems are also available for the large-scale production of anchorage-dependent cells.

Culture of Differentiated Cells

Differentiation is a process whereby cells slowly change their characteristics to become specialized cells with characteristic phenotypes. This process occurs *in vivo* during embryo development or during wound healing and leads to the formation of cells with specialized function (*differentiated*) such as neurons or muscle cells. Differentiation is also associated with normal cell replacement, as is necessary in the bloodstream. The undifferentiated precursors of this process are called stem cells.

Most stem cells or embryonic cells grow well in culture. However, as cells become more specialized (differentiated) they tend to lose their growth capabilities and this is reflected by poor growth in culture. For most cell types proliferation is incompatible with the expression of differentiated properties.

However, when some cells derived from a tissue are placed in culture there can be an apparent loss in differentiated properties in the surviving cell population. Some explanation can be offered for this:

1. Selective outgrowth of undifferentiated cell types. These include fibroblasts and epithelial cells that may be obtained from nongrowing animal tissue;
2. Adaptive response of cells to the culture media.

Tumour cells are, in most cases, undifferentiated and have good growth characteristics. However, there are also some differentiated tumour cells which have proved extremely valuable. For example neuroblastomas are fastgrowing tumour cells which have been used to study response effects with nerve growth factor.

Differentiated tumour cells retain the phenotypic characteristics of normal differentiated cells but are also able to grow in culture.

Although growth of most differentiated cells is poor, the following factors may allow some differentiated properties of normal cells to be maintained in culture.

1. *Hormones and growth factors*. There are a number of media formulations containing selective components that can maintain the differentiated state of specific cell types, for example keratinocytes, hepatocytes and nerve cells.
2. *Chemical agents*. Solvents such as *dimethyl sulfoxide* (DMSO) may allow the maintenance of a differentiated state by an effect on membrane fluidity.
3. *Cell interactions*. Contact between cells may allow the formation of gap junctions and allow metabolites to synchronize the expression of differentiation within a cell population. This may also play a part in the arrest of growth when a cell population has covered an available growth surface (defined as 'confluence').
4. *Interaction with the growth surface*. Collagen has been found to be essential for maintaining the polarity of hepatocytes in relation to the attachment surface. Cell polarity is governed by an asymmetrical distribution of ion currents (particularly Ca^{2+}). This allows one end of the cell to be functionally distinct from the other.

Some culture systems have been extremely valuable in investigating the metabolic changes that are associated with differentiation. However, growth in these cultures is either nonexistent or can be prolonged only for a short period (weeks).

Embryonic Stem Cells

These cells are capable of apparently unlimited growth but have the capacity, given the appropriate stimuli, to differentiate into any other cell type. Human embryonic stem cells were first isolated in 1998 by J. Thomson at the University of Wisconsin, who derived several cell lines and showed their capacity for growth for up to at least 300 population doublings. The cells were derived from the inner cell mass

(~30 cells) of a human blastocyte formed from several days growth of an embryo following *in vitro* fertilization. These embryonic stem cells have been shown to have several important properties.

1. Pluripotent. They have the capacity for differentiation into the cells of the three major tissue types (endoderm, mesoderm and ectoderm). This means that they have the potential to act as precursors for all cells of the body.
2. They can be propagated indefinitely in a non-differentiated state.
3. Directed differentiation. They can be induced to follow a specific pathway of differentiation, given the appropriate chemicals, growth factors or cell contact.
4. They are associated with specific cell markers, e.g. Oct-4 transcription factor and stage-specific embryonic antigen (SSEA).
5. They have a normal diploid karyotype.
6. The cells have a high activity level of telomerase. This tends to correlate with immortality in human cell lines.
7. If injected into immunocompromised mice they form a teratoma in which some of the major differentiated cell types can be distinguished.

Directed differentiation

If embryonic stem cells are allowed to clump then they form an embryoid body in which the cells begin to differentiate spontaneously. However, through directed differentiation the addition of specific growth factors may direct the cells down a specific pathway of change. The ability to direct such events is extremely useful for both studying developmental changes and also for use in cell therapy.

Adult Stem Cells

These are undifferentiated cells found among differentiated cells in a tissue or organ. Normally these cells can differentiate along a more limited pathway than embryonic stem cells to produce cells associated with the tissue. The cells serve to replace cells or repair tissue damage. Under certain conditions these cells may be induced into cell types other than those associated with the tissue from which they were derived. This is known as transdifferentiation or plasticity and is presently an active area of research.

Stem cells found in the bone marrow differentiate through the hematopoietic pathway to provide the extensive range of mature cell types. The hematopoietic stem cells are important for the continuous replacement of the cells found in the blood system. The hematopoietic pathway involves differentiation of cells through four stages - stem cells, early progenitor cells, progenitor cells and mature cells. The lymphoid progenitor cells produce the mature T-lymphocytes, B-lymphocytes and natural killer cells from progenitor cells that are stimulated to differentiate by various interleukins (IL-2, IL-3, IL: 6, IL-7). The myeloid lineage forms erythroid cells that can differentiate into monocytes, macrophages, neutrophils, eosinophils, basophils, megakaryocytes and erythrocytes. The formation of erythrocytes is stimulated by the glycoprotein, erythropoietin, the production of which occurs in the kidney and is enhanced by low oxygen levels (*hypoxia*). The steps of hematopoiesis are controlled by the microenvironment of the cells and secreted proteins (*cytokines*) from neighbouring cells (*stromal cells*) promote the differentiation process *in vivo*. Isolated progenitor cells can also be expanded outside the body (*ex vivo*). The early-stage hematopoietic progenitor cells are characterized by a surface antigen called CD34 by which the cells can be isolated. The most effective way of promoting the differentiation of these early progenitor hematopoietic cells in culture (*ex vivo*) is by cocultivation with stromal cells. Alternatively, conditioned stromal cell media or the addition of specific cocktails of recombinant allows expansion of the hematopoietic lineage. The use of a chemically defined medium containing recombinant cytokines is also the preferred method for therapeutic use because it ensures consistency and the absence of infectious agents. Many culture parameters affect the differentiation process including pH, osmolarity, temperature and media composition.

Fansformed Cells

Transformation has two different meanings in cell biology:

1. Expression of foreign genes in bacteria;
2. Change of animal cells from normal to infinite growth capacity.

In this section we are using the second definition. 'Normal' animal cells have a finite growth capacity but some cells acquire a capacity for infinite growth and such a population can be called an 'established' or continuous cell line. This requires cells to go through a process called transformation, which causes the cells to lose their sensitivity to the stimuli associated with growth control.

Transformed cells may also lose their anchorage-dependence and often show some chromosome fragmentation. This genetic state is referred to as aneuploidy, which means that there is a slight alteration from the normal diploid state. The transformed cells have a high capacity for growth in relatively simple growth medium and without the need for growth factors.

Carcinogenesis *in vivo* is analogous, but not identical to the transformation of cells *in vitro*. Not all transformed cells are malignant, a characteristic defined by the ability to form tumours in animals. However, all tumourderived cells grow continuously in culture. Examples include HeLa cells, which are derived from a cervical cancer and Namalwa cells, which derive from a human lymphoma. These cells are relatively easy to grow. They are robust and show good growth characteristics, which include a short doubling time and a low requirement for growth factors.

Cells can the transformed or 'immortalized' by a variety of techniques. This includes treatment with mutagens, viruses or oncogenes.

An oncogene is defined as a gene that induces the formation of tumourigenic cells. The first 'tumour' virus to be recognized was Rous sarcoma virus which is a chicken retrovirus described by Rous in 1911. The viral oncogenes (e.g. v-myc) are derived from cellular genes —the proto-oncogenes (e.g.: c-myc).

Infection by retrovinises is a particularly effective method of immortalizing cells. These retroviruses express activated oncogenes (e.g. *myc* and *ras*), which cause cell transformation. The retroviruses are also useful for incorporating recombinant DNA into animal cells. Alternatively, some cells can transform spontaneously in culture. This has been observed particularly with rodent cells and may be explained by the tendency of such cells to harbor endogenous viruses.

Cells from a Culture Collection

For many applications, cell lines may be obtained from cell culture collections ('cell banks'), which have a large selection of well-characterized cell lines. This is far easier than having to rely on primary animal tissue for establishing cultures. Samples of cell suspensions ($\sim 10^7$ cells/ml) are offered for sale by the cell collections in frozen vials that can be transported in dry ice. On arrival the cells should be thawed and inoculated immediately into culture media or stored in liquid nitrogen.

All these cells grow well in culture when provided with an appropriate growth medium. The full history of each cell line is available from the collection catalogs as well as a list of references indicating previous use of the cells.

Table 3.1. Examples of common cell lines obtainable from culture collections.

Cell line	*Origin*	*Cell type*	*Comment*
BHK	Baby hamster kidney	Fibroblast	Cells are anchorage-dependent but can be induced into suspension; used for vaccine production
CHO	Chinese hamster ovary	Epithelial	Cells will attach to a surface if available but will also grow in suspension; used extensively for genetic engineering
HeLa	Human cervical carcinoma	Epithelial	Fast-growing human cancer cell isolated in the 1950s
L	Mouse connective tissue	Fibroblast	Many culture techniques developed from the 1950s were based on this tumor cell line
L6	Rat skeletal muscle	Myoblast	Can be used to demonstrate the differentiation of a muscle cell
MDCK	(Madin Darby) carcine kidney	Epithelial	Anchorage-dependent cells with good growth characteristics; used for veterinary vaccine production
MRC-5	Human embryonic lung	Fibroblast	Finite lifespan; normal cells; used for human vaccine production
MPC-II	Mouse myeloma	Lymphoblast	Derived from a mouse tumor secretes immunoglobulin
Namalwa	Human lymphoma	Lymphoblast	Derived from cells from a human suffering from Burkitt's lymphoma; used for alpha-interferon production
NB4IA3	Mouse neuroblastoma	Neuronal	Tumor cells with good growth rate. Cells have nerve cell characteristics including a response to nerve growth factor
3T3	Mouse connective tissue	Fibroblast	Vigorous growth in suspension; cells used widely in the development of cell culture techniques.
WI-38	Human-embryonic lung	Fibroblast	Finite lifespan; normal cells; used for human vaccine production
Vero	African green money kidney	Fibroblast	An established cell line capable of continuous growth but with many normal diploid characteristics; used for human vaccine production

The largest and most well-known international animal cell culture collections are given below. These collections contain well over 3000 well-characterized cell lines.

1. The American Type Culture Collection (ATCC)
2. The European Collection of Animal Cell Culture (ECACC).

PROTOCOLS

Preparing a Primary Cell Culture of Chick Embryo Fibroblasts

Equipment

Laminar flow cabinet

Forceps

Screen cup + 220 μm mesh (Sigma)

Materials

10 embryonated eggs

Hank's balanced salt solution (calcium- and magnesium-free)

Glass beads (1 mm diameter: Sigma)

0.2% (v/v) trypsln (1:1250) in Hank's balanced salt solution

Dulbecco's modification of Eagle's medium (DMEM)

Fetal calf serum (FCS)

Phosphate buffered saline (PBS)

Protocol

1. Clean 10 fertile 10-day-old incubated eggs with 70% ethanol.
2. Open the broad end of each egg with sterile scissors and carefully remove a piece of shell with the white membrane underneath.
3. Remove the embryos by sterile forceps, discard each head and transfer the trunk to a Petri dish containing 10 ml of sterile Hank's balanced salt solution (Ca- and Mg-free).
4. Wash x2 with the salt solution to remove erythrocytes.
5. Remove and discard the internal organs with forceps.
6. Wash the remaining carcass x 3 with salt solution.
7. Transfer to a dry sterile tissue culture dish and chop into small pieces with sterile scissors.
8. Transfer to a flask containing 0.5 g glass beads (1 mm diameter) suspended in 1 ml PBS.
9. Add 20 ml 0.2% (v/v) trypsin (1:1250) in Hank's balanced salt solution and swirl the flask gently for 30 seconds.
10. Allow the larger fragments to sediment, then remove and discard the supernatant containing the erythrocytes.

11. Add a further 20 ml of trypsin, agitate for 30 seconds and transer supernatant to a sterile centrifuge tube. Repeat this trypsin extraction x5.
12. Centrifuge the tube at 300 g for 5 min to pellet the cells. Re-suspend in Hank's balanced salt solution containing 5% fetal calf serum.
13. Pour the cell suspension through a sterile 220 μm mesh screen cup to remove all debris.
14. Pellet the cells from the suspension and re-suspend in growth medium (DMEM + 10% FCS).

Notes

This entire procedure should be carried out in a laminar flow cabinet under sterile conditions. The process combines proteolytic digestion with mechanical agitation to cause individual cells to disperse into a liquid suspension. The addition of a serum-supplemented medium after trypsinization is important to inhibit further activity of the trypsin. Digestion of the tissue may be aided by the proteolytic enzyme collagenase and by the addition of chelating agents such as Versene or ethylene-diaminetetra-acetic acid (EDTA).

Isolation of Lymphocytes from a Blood Sample

Materials

Diluted blood: Add an equal volume of balanced salt solution to anticoagulanttreated blood

Balanced salt solution (120 mM NaCl,0.55 mM glucose, 5 μM $CaCl_2.2H_20$, 0.98 μM $MgCl_2.6H_20$, 0.54 mM KCI, 14.5 mM Tris)

Protocol

1. Add 3 ml Ficoll-Paque to a centrifuge tube.
2. Layer 4 ml of diluted blood sample on to the Ficoll-Paque.
3. Centrifuge at 400 g for 30 minutes at room temperature.
4. Remove the upper layer (plasma) with a Pasteur pipette, leaving the lymphocyte layer undisturbed.
5. Transfer the lymphocyte layer into another centrifuge tube using a minimal volume.
6. Add three volumes of balanced salt solution.
7. Aspirate with a pipette to ensure adequate suspension of the cells in the liquid.
8. Centrifuge at 100 g for 10 min and discard the supernatant.
9. Re-suspend the cell pellet and repeat steps 6 to 8 above.

Notes

The upper layer following the initial centrifugation contains blood plasma and is a good source of platelets. Washing the lymphocytes in salt solution ensures removal of the platelets. Granulocytes and erythrocytes form the lower layer after centrifugation. Transferring a minimal volume of the lymphocyte band ensures minimal granulocyte contamination. Typically this procedure recovers up to 80% of the lymphocytes from the original blood sample.

4

NATURE OF CULTURE

In this chapter the characteristics of simple batch cultures are compared to alternative modes of culture that allow continuous or semicontinuous feeding. These continuous culture systems are valuable for obtaining high cell yields even from low-volume cultures. This enables product yields to be increased by higher cell densities in culture without necessarily a scale-up in volume. This is called 'process intensification. Cell immobilization into suitable inert carrier beads can help in obtaining these high yields, particularly for anchorage-dependent cells and is explained in this chapter.

BATCH CULTURE

In the previous chapter different bioreactor types were considered from the simple stirred tank reactors to the more complex packed-bed or hollow-fiber systems. These bioreactors can be classified in terms of their mixing characteristics as homogeneous or heterogeneous. In the stirred tank reactor, homogeneous mixing of all culture components is maintained. A cellular substrate introduced into the bioreactor will blend immediately into the culture and will gradually decrease in concentration as cells grow.

On the other hand, the packed-bed and hollow-fiber systems can be characterized as heterogeneous. A substrate introduced into the inlet will flow as a single band through the tubular bioreactor without back mixing. This is called *'plug-flow'*. The graph shows the change in concentration of a substrate at the inlet of such a bioreactor. The concentration will decrease along the length of the bioreactor. For better utilization of substrates such heterogeneous systems are normally fitted with a loop so that the culture medium is continuously recycled

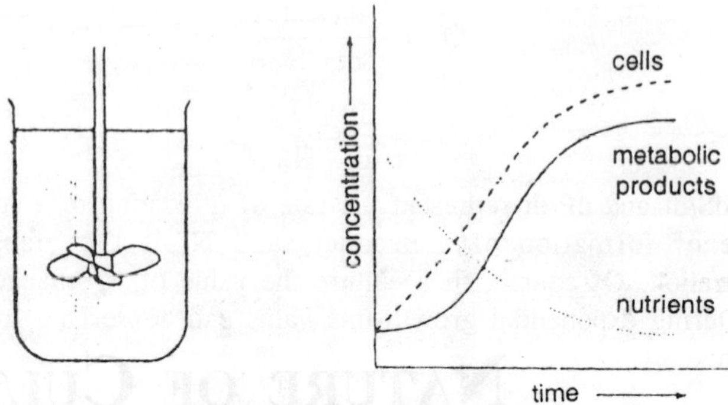

Fig. 4.1. Kinetics of a batch culture.

over the growing cells. This is called a quasi-homogeneous system in which the concentration of any substrate will decrease in a series of steps. Here the kinetics of substrate utilization approximates what happens in a homogeneous culture.

All the above are examples of closed or batch cultures. The cells are inoculated and the culture is left for several days until the final density is reached. In a closed system nothing is added or removed during the culture, apart from small volume samples for analysis. However, it is important to realize that the cells are exposed to a constantly changing environment of substrates that are removed from the medium and products that are secreted from the fells. Any nutrient will gradually decrease from an initially high concentration to a low concentration following growth of the cells. These changes lead to a gradually deteriorating environment for cell growth, which is likely to cease at around 10^2 cells/ml because of one or more of the following factors:

1. Depletion of an essential nutrient;
2. Accumulation of an inhibitor;
3. A complete cover of the available growth surface (for anchorage dependent cells).

Media components are consumed by the cells at variable rates, glutamine and glucose being the substrates consumed most rapidly during growth. Their utilization leads to the gradual formation of ammonia and lactate as metabolic by-products. Substrate consumption or product formation is usually expressed as a specific rate as pmol/cell per hour or μmol/10^6 cells per hour. This measurement of specific substrate uptake (Q_s) or product formation (Q_p) is widely used in the analysis of media changes during batch culture.

$$Q_S = -\frac{dS}{dt} \cdot \frac{1}{N_V}$$

$$Q_P = \frac{dP}{dt} \cdot \frac{1}{N_V}$$

where dS/dt and dP/dt represent the rate of utilization of a substrate or rate of formation of a product, and N_v is the viable cell concentration. Of course, in a culture the value of N_v changes over time. During exponential growth this value can be derived from the equation:

$$Q_S = \frac{\Delta S}{T} \cdot \frac{\ln N - \ln N_0}{N - N_0}$$

where ΔS is the change in concentration of the substrate of the time period T, and N_0 and N are the initial and final cell concentrations, respectively.

This equation is applicable during the exponential growth phase only. However, over the whole culture where the growth goes through several phases, the equation becomes:

$$Q_S - \frac{\Delta S}{\int_0^1 N_V dt}$$

The integral represents the area under the growth curve (N vs time) from time zero to each time point. If such a calculation is made for each time point a series of integral values is generated. If N represents viable cell concentrations then the integrals are often referred to as viability indices. The value of Q_s (or Q_p) can be determined from the slope of the plot of substrate (or product) concentration and the integral values. A single straight line indicates a constant rate of substrate uptake per cell throughout the culture. In some cases distinct phases of uptake or production rates may be observed. Typical data from a batch culture in which cells grow to a maximum density of 10^6 cells/ml, as a substrate at an initial concentration of 25 mM is utilized. Transformation of these data suggests two phases of uptake during culture with a high Q_s during the cell growth phase and a lower Q_s during the stationary and decline phases. The same type of plot can be used to determine production rates of metabolic by-products such as lactate and ammonia.

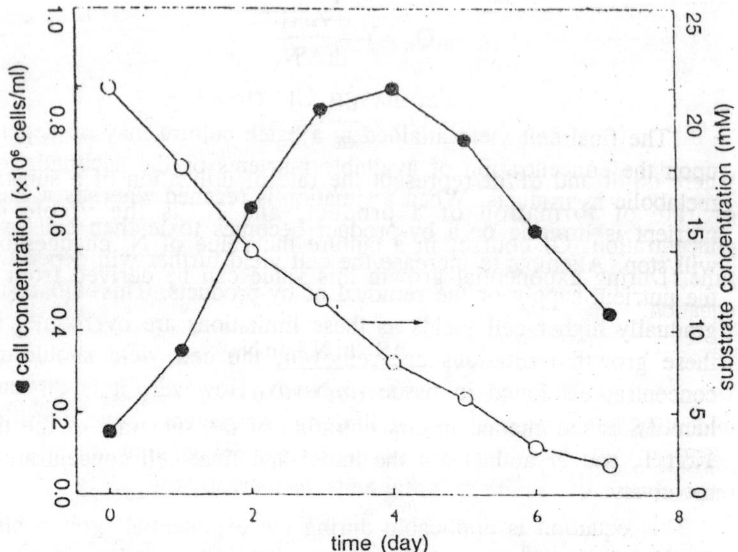

Fig. 4.2. Cell growth and substrate concentration.

Another useful measure is the growth yield ($Y_{N,S}$) which is the measure of the number of cells generated following the consumption of a millimole of substrate.

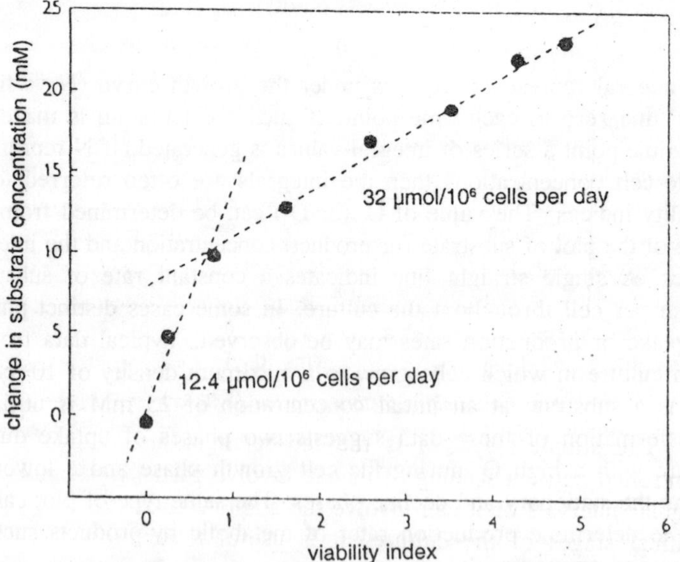

Fig. 4.3. Viability index vs substrate utilization.

$$Y = \frac{\Delta N}{\Delta S}$$

Fed-batch Culture

The final cell yield attained in a batch culture may be dependent upon the concentration of available nutrients or the accumulation of metabolic byproducts. When a situation is reached whereby a specific nutrient is limiting or a by-product becomes toxic then cell growth will stop. Attempts to increase the cell yield further will depend upon the nutrient supply or the removal of by-products. This will result in gradually higher cell yields as these limitations are overcome. If all these growth limitations are removed, the cell yield should reach concentrations found in tissue (*in vivo*). However, it is difficult to identify the sequential list of limiting parameters above densities of 10^7 cells/ml. In an open system cells are supplied with nutrients during the course of culture in an attempt to increase the cell yield.

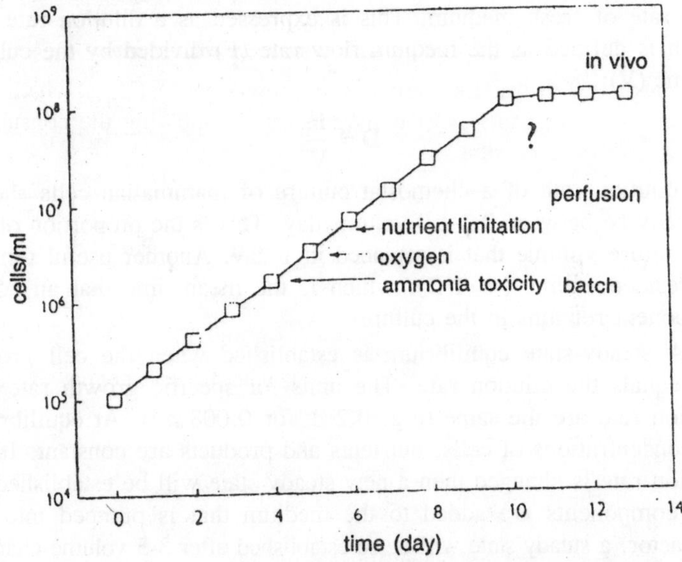

Fig. 4.4. Limiting factors associated with cell growth.

The simplest system is the fed-batch culture which involves controlled nutrient feeding. This can involve partial media changes at regular intervals or it can involve the addition of specific nutrients at critical stages of the culture. It is useful to identify the limiting nutrients before setting up a fed-batch culture. This can lead to the addition of a concentrate of specific amino acids or glucose at daily intervals in an attempt to increase the final cell yield. However, despite such feeding, metabolic products will still accumulate eventually to growth-inhibitory concentrations.

Continuous Culture

This term is applied to an open system with a continuous feed of medium and removal of 'spent' medium. Cell growth continues for longer periods in a continuous culture compared to a batch culture. There are two main types of continuous culture:

1. Chemostat culture: where the cells are removed continuously from the fermenter;
2. Perfusion culture: where the cells are retained in the fermenter.

Chemostat Culture

In this system spent medium and cells are removed continuously through an outlet tube. The state of the culture is dependent upon the flow rate of fresh medium. This is expressed as a dilution rate (D) which is defined as the medium flow rate (F) divided by the culture volume (V):

$$D = \frac{F}{V}$$

The dilution rate of a chemostat culture of mammalian cells should normally be between 0.2-1.0 volume/day. This is the proportion of the cell culture volume that is replaced in 1 day. Another useful term is the residence time (= 1/D), which is the mean time that an added component remains in the culture.

A steady-state equilibrium is established when the cell growth rate equals the dilution rate. The units for specific growth rate and dilution rate are the same (e.g. 0.2 d^{-1} or 0.008 h^{-1}). At equilibrium the concentrations of cells, nutrients and products are constant. If the dilution rate is changed then a new steady state will be established. If new components are added to the medium that is pumped into the bioreactor, a steady state will be re-established after 3-5 volume changes of the culture.

The chemostat is a self-regulating system in which a temporary decrease in cell concentration will cause a corresponding increase in growth rate, and an increase in cell concentration has the reverse effect. Thus, the steady-state conditions are always restored, although each new equilibrium will take 3-5 volume changes to be established. In the chemostat the specific growth rate (μ) of the cell population is maintained at a level below a maximum value (μ_{max}) by the concentration of a limiting substrate (S). This is defined by the Monod equation:

$$\mu = \mu_{max} \cdot \frac{S}{K_S + S}$$

where S is the variable concentration of the limiting substrate, and K_s is a constant related to the affinity of the cell-substrate interaction.

When S is significantly higher than K_s then the nutrients are in excess and the growth rate is at the maximal value. As the substrate concentration declines to the value of K_s then so does the growth rate. The saturation constant (K_s) from the Monod equation represents the substrate concentration at which the specific growth rate is half the maximum rate ($\mu = \mu_{max}/2$). This parallels the relationship between substrates and enzymes as expressed in the well-known Michaelis-Menten equation. In a continuous culture the growth rate will increase as the concentration of the limiting substrate is increased, until a critical value is reached when cell growth continues at its maximum rate. If the dilution rate is increased beyond this point, then the cells will be washed out of the culture. This occurs when the medium is pumped out of the culture faster than the cells can grow.

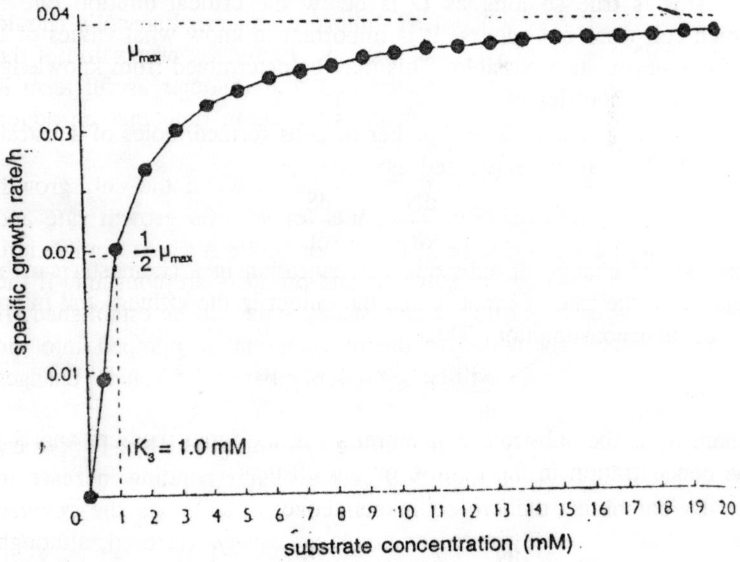

Fig. 4.5. The effect of substrate concentration on specific growth rate.

Mammalian cells can be grown in a chemostat culture in which the glucose concentration is limiting, although in many cases the limiting substrate may be unknown. The following are the kinetic equations that define a chemostat culture. The increase in cell number:

$$\frac{dN}{dt} = \mu.N - D.N$$

where μ.N represents an increase in cell number from growth and D.N represents a decrease in cell number from the effluent of the culture.

If the specific growth rate is greater than the dilution rate ($\mu >$ D) then dN/dt is positive and the concentration of cells in the culture will increase.

If the specific growth rate is lower than the dilution rate ($\mu <$ D) then dN/dt is negative and the concentration of cells in the culture will decrease. If this is maintained then the cell number will eventually decrease to zero indicating complete wash-out from the culture.

At steady state:

$$\frac{dN}{dt} = 0$$

at which μ = D.

This is true so long as D is below the critical dilution rate at which cell wash-out occurs. It is important to know what values of D make a steady state possible. This may be determined from knowledge of the yield coefficient:

Yield coefficient, Y = number of cells formed/moles of substrate utilized This can be expressed as:

$$\frac{dN}{dt} = Y \cdot \frac{dS}{dt}$$

The rate of change of substrate concentration in a chemostat can be equated to the rate of input minus the output in the effluent and minus the cellular consumption. Thus:

$$\frac{dS}{dt} = D.S_R - D.S - \frac{\mu N}{Y}$$

where S_R is the substrate concentration in the input medium and S is the concentration in the culture or the effluent.

By combining the two equations above:

$$\frac{dN}{dt} = N\left(\mu_{max} \cdot \frac{S}{K_S + S} - D\right)$$

and

$$\frac{dS}{dt} = D.(S_R - S) - \mu_{max} \cdot \frac{S}{K_S + S} \cdot \frac{N}{Y}$$

If S_R and D are held constant and D does not exceed a critical value, then there are unique values of N and S for which both dN/dt = 0 and dS/dt = 0. These are the steady-state values of the cell concentration (N') and substrate concentration in the bioreactor (S'):

$$S' = K_S \cdot \left(\frac{D}{\mu_{max} - D} \right)$$

$$N' = Y \cdot (S_R - S) = Y \cdot \left(S_R - K_S \cdot \frac{D}{\mu_{max} - D} \right)$$

Over most of the range of possible dilution rate values, the steady-state concentration of the limiting substrate is very low. This indicates that it is almost completely consumed by the cells.

When D approaches the critical dilution rate (D_c) at which wash-out occurs, then unused substrate appears in the culture.

At D_c the substrate approaches the maximum value ($S = S_R$). Therefore

$$D_c = \mu_{max} \cdot \frac{S_R}{K_S + S_R}$$

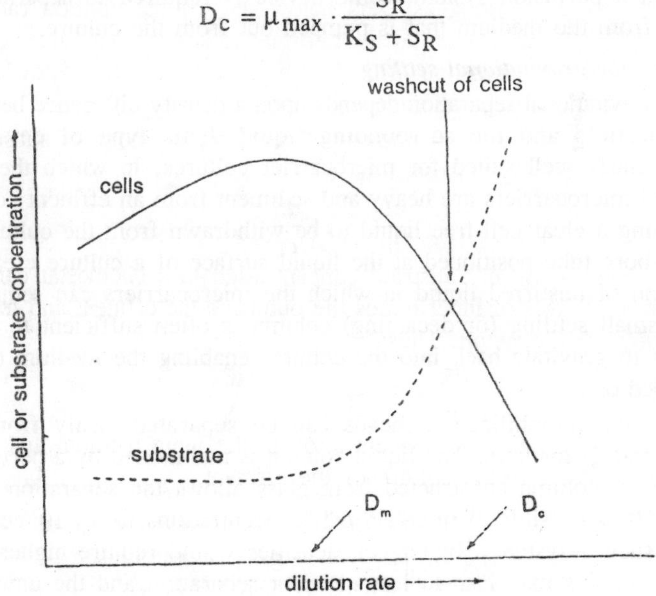

Fig. 4.6. The effect of dilution rate on cell and substrate concentration in a chemostat.

The cell and product output from the chemostat increases with D × N, which is the dilution rate times the cell concentration. At D_c there is complete wash-out of cells. However at a lower value than this (= D_m) the maximum cellular output and production occurs. This value corresponds to the maximum growth rate (μ_{max}) of the cells.

Perfusion

An alternative type of open system involves medium perfusion. Here, the medium is pumped continuously through the culture, but the cells are retained in the bioreactor. This type of process is becoming increasingly popular for large-scale production. The continuous supply of nutrients and removal of waste products promotes cell growth to a greater extent than in a batch culture and allows a high cell density to be attained. Typically, final cell densities of $>10^7$ cells/ml will occur although this is dependent upon the perfusion rate. The advantages of the perfusion mode include high productivity from a high cell density and a low residence time of the product in the culture environment. he low residence time of a recombinant protein has the advantage of reduced risk from any protease and sialidase activity that may be released by the cells.

Cell separators

In a perfusion system, some device is required to separate the cells from the medium that is pumped out from the culture.

Open tube/gravitational settling

Gravitational separation depends upon a density difference between the particle and the surrounding liquid. This type of device is particularly well suited for microcarrier cultures, in which the cell-loaded microcarriers are heavy and sediment from an effluent column allowing a clear cell-free liquid to be withdrawn from the culture. A wide-bore tube positioned at the liquid surface of a culture creates a column of unstirred liquid in which the microcarriers can sediment. The small settling (or decanting) column is often sufficient to allow heads to gravitate back into the culture, enabling the medium to, be pumped out.

Cells immobilized in beads can be separated easily from the surrounding medium. The liquid column is maintained by a peristaltic pump. A column constructed from glass allows the separation to be seen. It is possible to maintain cell concentrations up to 10^7 cells/ml with such a system. Higher cell densities would require higher flow rates but this may lead to less efficient separation and the unwanted removal of cells from the bioreactor. Unfortunately, this type of column

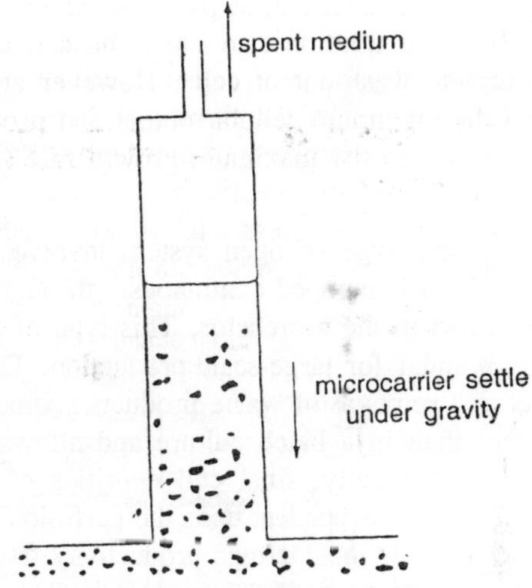

Fig. 4.7. Column for gravitational settling of microcarriers.

separator will not operate efficiently for a suspension culture because the density of cells is not sufficiently different from the surrounding medium to allow separation. However, the principle of gravitational separation can be applied in a lamella settler, in which the cells are separated in a series of closely packed parallel plates, normally inclined at a 45° angle. The key to this device is the large area offered for separation which allows cells to move down the lower surface of the separation space. Separation is aided by coating the plates with a material to prevent cell adhesion and also by continuous vibration.

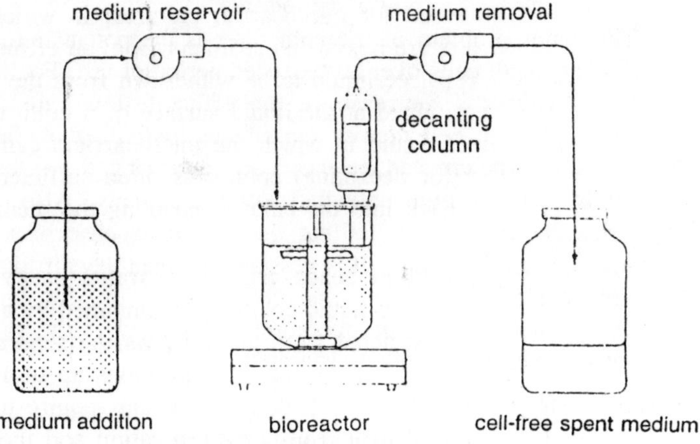

Fig. 4.8. Continuous culture with a decanting column.

Spin filter

This device involves a rotating mesh cage to allow medium flow but prevent cell wash-out. The cell-free effluent is withdrawn from within the filter. Rotation of the filter is required to prevent clogging. The device may be constructed from stainless steel with a variable mesh size. A mesh size of 5-15 microns is suitable for cell separation from a suspension culture whereas a higher mesh size of 70-150 microns is more suitable for a microcarrier culture.

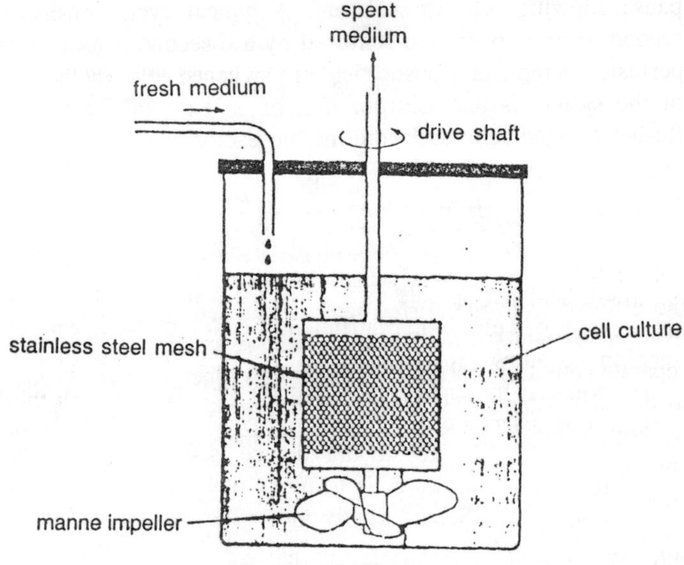

Fig. 4.9. Spin filter.

The major problem of the spin filter is the potential for fouling and blockage with cells over an extended period of use. Fouling of the spin filter causes a decrease in the effluent flow rate from the bioreactor. A high cell density contributes to fouling of the screen, although this problem can be eased by increasing the filter surface area or by using a high enough rotational velocity. Also, a higher mesh size could alleviate fouling but at the expense of a reduced efficiency of cell separation. It is possible to mount the stirring impeller and the spin filter on the same shaft. However, for large-scale bioreactors there is greater operational flexibility if the rotation of the two devices is controlled separately.

Acoustic cell filtration

In this device the effluent from a culture is withdrawn through a resonator chamber, in which the liquid is subjected to an ultrasonic acoustic field at a frequency of about 2 MHz. A typical configuration of this system is provided by Applikon with their Biosep device. The standing sound wave, which is perpendicular to the outward flow of liquid, causes cells to aggregate and sediment back into the bioreactor. In its simplest operation, the perfusion pump runs continuously. However, there are some operational refinements that may improve the performance and efficiency of separation. One method is to operate the perfusion pump in a discontinuous cycle, which can incorporate a pause allowing cells to sediment. A typical cycle consists of a 20-second forward operation followed by a 6-second pause, in which the perfusion pump and acoustic field are switched off. Another refinement of the system is the incorporation of an external loop pump which flushes trapped cells back into the bioreactor against the main effluent flow.

Tangential flow filtration/loop reactor

The principle of this separation is that the culture is fed through hollow fibers, the walls of which are permeable. In this loop bioreactor the effluent from the stirred tank is pumped continuously through an external tangential flow filtration unit. A small ultrafiltration unit can separate cells from medium in the hollow fibers. The cells are returned to the stirred tank. The flow rate through the filtration unit is limited by high shear rates that can damage the cells. Membrane blockage can be a problem but is preventable by periodic back flushing and coating the inner surface with polyethylene glycol that prevents the build-up of protein. To scale-up the separation, a large number of short filtration units should be used in parallel.

CELL IMMOBILIZATION

The immobilization of cells on or inside particles suspended in culture medium offers an alternative mode of culture. Immobilization can involve the attachment of cells to a solid surface—which is the only way anchorage-dependent cells can grow. Alternatively, it can mean the entrapment of cells in small beads. In this section we will look at the properties of various immobilization systems. These immobilized systems are ideal for use with perfusion cultures.

Microcarriers for anchorage-dependent cells

At the laboratory scale, anchorage-dependent cells can be grown in Petri dishes, T-flasks or roller bottles. However, these are limited in terms of the available surface area for growth. In order to increase cell yield or to design a process for large-scale production involving

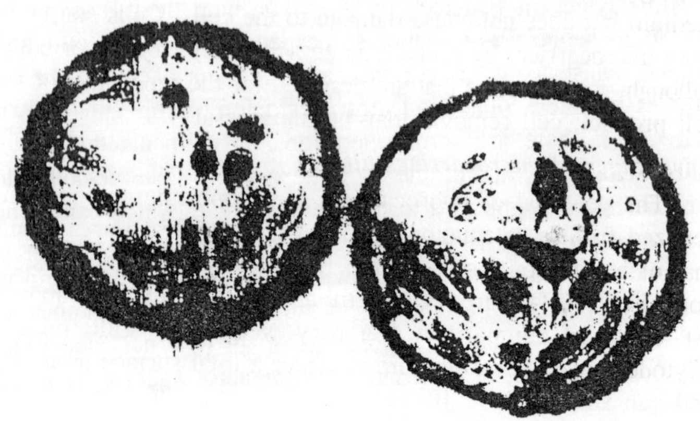

Fig. 4.10. A confluent layer of BHK cells covers two microcarriers.

anchorage-dependent cells, a system offering a high growth surface to volume ratio is required. Microcarrier cultures are designed to provide a pseudo-suspension culture for anchorage-dependent cells.

Microcarriers are microscopic particles (normally with a diameter of about 200 μm) that are easily maintained in suspension in liquid medium. The suspension of microcarriers is maintained in a spinner flask which contains a magnetic flea and a paddle suspended from a stirring shaft. The flask can be 100-500 ml in volume and should be placed on a specialized microcarrier stirring platform to allow a relatively slow and controlled stirring speed of around 40 rpm. These cultures are also suitable for scale-up and can be operated in simple stirred tanks in production processes.

The major characteristics required of microcarriers are as follows:

1. Small, to maximize the available surface area for cell growth;
2. Light, to allow easy suspension in culture medium;
3. Transparent, to allow easy observation of cell attachment and growth;
4. Charged, to allow cell attachment onto the surface.

Microcarriers have been made from various materials including dextran, collagen and plastic. The surface is charged so as to ensure cell attachment. The density of the electrostatic charge on the microcarrier surface is critical to allow cell attachment and growth. The cells will then grow over the outer surface area until a monolayer is formed. Each microcarrier will accommodate 100-200 cells. It is important at inoculation that there is an even distribution of cells on the available microcarriers so that there are no 'empty' microcarriers at the end of the growth period. Also, it is important to ensure a stirring rate that allows the microcarriers to be suspended in the culture medium but does not cause damage to the cells. Cells can be detached from microcarriers by proteolytic enzymes (e.g. trypsin or collagenase) although, for processes that are designed for the production of a secreted cell product, cell removal may be unnecessary.

Advantages of microcarrier cultures

The potential of microcarriers for supporting cell growth can be realized from a calculation of the available surface area compared to that of a traditional system, the Roux bottle. A standard-sized Roux bottle has a surface area of 0.02 m^2 and can accommodate approximately 10^8 cells. In comparison 1 g (dry weight) of solid microcarriers (Cytodex) in 1 liter of culture provides a total surface area of 0.64 m^2 and can support 2×10^9 cells (equivalent to 20 Roux bottles). The handling operations involved in a 1-liter microcarrier culture are considerably simpler than those of 20 Roux bottle cultures, yet an equivalent cell yield is obtained. The advantages of a unit system become even clearer as the culture is scaled tip to higher volume.

A key factor in obtaining a high cell yield in a microcarrier culture is to ensure that each bead accepts a viable cell at inoculation. This ensures full colonization of the available surface and limits the number of unoccupied beads at the end of the culture. Attachment of cells to beads has been shown to follow a Poisson distribution:

$$P = e^{-\lambda} \cdot \frac{\lambda^n}{n}$$

where P = probability of a specific number (n) of cell hits per bead, and λ = ratio of cells to beads in culture at inoculation.

The most useful result obtained from such an analysis is that a critical cell-to-bead inoculation ratio is necessary to ensure a negligible proportion of unoccupied microcarriers at the end of the culture. For most viable cell lines this critical ratio is >7. This ensures that the proportion of unoccupied microcarriers is $<5\%$ and the use of the available surface area is maximized.

A typical microcarrier culture is established with 1-3 g (dry weight) of microcarriers per liter. This allows a final cell density of around 2×10^6 *cells/ml* to be obtained from an inoculum density of 2×10^5 cells/ml. Higher microcarrier concentrations can be used (10-15 g/liter) but the disadvantages are the increasing difficulty of adequate culture agitation and the high cell inoculum densities required to enable correspondingly high yields.

Porous microcarriers

The original dextran microcarriers (e.g. Cytodex) are microporous. The pore size is not sufficient to allow cells to colonize the interior of the beads. However, macroporous carriers such as Cultispher are spherical gelatin microcarriers containing open channels large enough for cells to enter and continue growing within each bead. These have the advantage of increased surface area for cell attachment and an interior environment that may protect cells against adverse shear forces. Cells can attach to the beads or they may become trapped within the channels of the bead. This allows much higher cell-bead ratios than with solid microcarriers.

Porous microcarriers have been designed with various sizes and weights and are suitable for use in stirred tank or fluidized bed bioreactors. Some porous microcarriers have the capacity to accommodate up to 2000 cells per bead.

Choice of microcarriers

There is a wide selection of commercially available microcarrier types which differ in terms of the basic supporting matrix and the surface-coating material. The choice of microcarrier will depend upon the cell line used and the purpose or objectives of the culture. The most commonly used for surface attachment are Cytodex 1, 2 and 3. The Cytodex 1 is a cross-linked dextran matrix with positively charged N,N-diethylaminoethyl (DEAE) groups throughout the matrix. This type of microcarrier is suitable for the culture of a range of anchorage-dependent cells and is the most widely used. For relatively robust established cell lines, the charged-dextran microcarriers (such as Cytodex 1) have proved efficient and reliable. Cytodex 2 has a surface charge only and may reduce potential binding of cell products. Cytodex 3 has a collagen layer coupled to the surface and may enhance initial attachment of some cells, particularly those from primary tissue. Cultispher G is a macroporous gelatin and allows the cells to colonize the interior allowing higher cell numbers per bead. Under a microscope, the cells can easily be seen on the microcarrier surface. Staining with crystal violet is particularly useful as this highlights the nuclei against the transparent background of the microcarrier surface. Dextran-based microcarriers have been used extensively for the large-scale production of biologicals such as vaccines that are produced from anchorage-dependent cell lines.

Bead-to-bead transfer is an important operation for the scale-up of microcarrier cultures. This occurs by the addition of new microcarriers into an existing culture in which cells begin to colonize the new microcarriers. For this to occur efficiently, cells must detach from a confluent microcarrier surface and this will depend upon how firmly the cells are bound to the surface. For some cell lines this can occur in a low calcium medium or by a change of pH. For other cases a cocktail of trypsin and EDTA may be required.

If it is necessary to isolate high cell yields of fragile cell lines then microcarriers other than charged-dextran may be considered. Glass-coated or collagen-coated microcarriers are particularly suitable because detachment can be achieved efficiently and without significant cell damage by either trypsin or collagenase treatment. Detached cells can then be separated from microcarriers by sieving through a nylon mesh of appropriate size.

Gelatin based or collagen-coated microcarriers may be a particularly good choice for the culture of primary cells. Such cells have fastidious growth requirements and attachment to a natural extracellular matrix may be advantageous. Microcarriers composed entirely of gelatin can be completely dissolved by proteolytic enzymes leaving the cells in suspension.

Immobilization of Nonanchorage-dependent Cells

The immobilization of nonanchorage-dependent cells into small beads can have many advantages such as:

- protection against mechanical stress;
- ease of continuous operation;
- isolation of products.

Two systems have been found applicable:

- encapsulation;
- entrapment.

Cell entrapment

This can be accomplished in a variety of large polymers. Some of these are polysaccharides, with agarose being particularly well suited. The entrapment process involves mixing and agitating a suspension of cells in warm agarose with a hydrophobic liquid such as paraffin oil at a lower temperature. This causes the formation of small solid beads of agarose (100-200 μm) containing the suspended cells. The mechanical strength of the beads can be varied by using different agarose concentrations from 0.5 to 5%. Agarose is highly porous and will allow the free diffusion of high-molecular-weight substances through the matrix.

Growth of the entrapped cells in the beads can continue to high densities in bioreactors with high stirring speeds and rates of aeration. At high densities, some cell leakage can occur from the beads, but this may be prevented by careful growth control. Secreted cellular products accumulate in the medium which can then be easily separated from the entrapped cells—the system being ideally suited for continuous perfusion.

Encapsulation

Encapsulation is a modification of cell entrapment with cells enclosed in a semipermeable membrane which retains the cells and large secreted molecules but allows the free diffusion of small molecules. Spherical solid beads of calcium alginate can be formed by allowing a suspension of cells in sodium alginate to drip into a calcium chloride solution. An outer semipermeable membrane can be created by the addition of a basic polymer (e.g. polylysine) and the calcium alginate interior of the bead can then be re-liquefied by treatment with sodium citrate. The encapsulated cells are well protected against mechanical forces and can accumulate to high densities within the membrane. High-molecularweight products are also retained within the membrane. This idea has been applied commercially to the growth of hybridomas. The secreted monoclonal antibodies accumulate to a high concentration within the bead matrix. Extraction of Mabs from beads of encapsulated cells is relatively easy because of the high concentration.

Protocol

Establishment and Monitoring of a Microcarrier Culture

Materials

Spinner flask 250 ml with a working culture volume of 100 ml
Microcarriers: Cytodex 1, 2 or 3 or Cultispher *G* (Hyclone) 10 g
Dulbecco's phosphate buffered saline (D-PBS)
D-PBS/EDTA: D-PBS + 0.02% EDTA
Trypsin reagent: 0.25% w/v trypsin + 0.02% w/v EDTA
Trypan blue reagent: 0.5% w/v trypan blue in D-PBS
Crystal violet reagent 0.2 M citric acid, 0.2% w/v crystal violet and 25 w/v Triton X-100

Method to establish a culture

1. Suspend dry microcarriers (10 g) in 500 ml D-PBS.
2. Wash with D-PBS and autoclave.

3. Allow the microcarriers to settle, remove excess D-PBS by decanting and add culture medium to the microcarriers.
4. Inoculate 100 ml culture medium containing 5 mg/ml Cytodex microcarriers with 2 × 10^5 cells/ml in a spinner flask.
5. Place the spinner flask on a stirrer platform positioned inside a CO_2 incubator and maintain a stirring rate of 50 rpm.
6. Take a sample every day by withdrawing 1 ml of culture and insert into an Eppendorf tube (1.8 ml) for cell counting.
7. Allow the microcarriers to settle and withdraw the supernatant for analysis of medium components.

Counting method 1: trypan blue counting

8. Wash the microcarrier/cell samples with PBS/EDTA and then treat with 1 ml trypsin reagent.
9. Incubate for 10 min at 37°C.
10. Remove the supernatant by Pasteur pipette and add an equal volume of trypan blue (0.5%) to each sample prior to counting cells by hemocytometer.

Counting method 2: crystal violet counting

11. Add an equal volume of crystal violet reagent to the microcarrier/cell pellet.
12. Incubate for 90 min at 37°C.
13. Count the stained nuclei that are released from the microcarriers by herocytometer counting.

Notes

The microcarriers are supplied as a dry powder and are reconstituted in aqueous buffer. Cytodex 1 consists of 4.5 × 10^6 beads/g dry weight. Cultispher G consists of 0.9 × 10^6 beeads/g dry weight. The dry microcarriers may swell in liquid. The spinner flasks should be siliconized before use to prevent microcarriers sticking on to the glass surface. A siliconizing fluid (dimethyldichlorosilane dissolved in an organic solvent) is used to wet the inner surface of a spinner flask prior to washing and autoclaving. One coating of the siliconizing agent is sufficient for several cycles of use of a spinner flask. Commercially available siliconizing fluids include Sigmacote and Repelcote.

5

SCALING-UP CULTURE

Small volume cultures are preferable for experiments requiring a large number of replicates. This might be necessary, for example, in the determination of the concentration-dependent effects of substances that promote or inhibit cell growth. Multi-well plates (e.g. 24-well plates) can accommodate culture volumes of 2-3 ml which are adequate for single determinations of cell concentration which may be taken after a suitable growth period of a few days. If an experiment requires multiple samples to be taken over time to observe cell growth and to perform substrate or product assays, then culture volumes up to 100 ml are usually adequate. Such experiments would normally be conducted in T-flasks or spinner flasks in a temperature-controlled CO_2 incubator. Small volume cultures have a large surface area/volume ratio. 'Ibis allows easy diffusion of gases through the culture surface.

Culture scale-up is required to produce substantial quantities of a cell product, such as a virus or a glycoprotein, or to obtain a sufficient quantity of cells for metabolite or enzyme extraction. There are two approaches to scale-up:

1. A multiple process would involve handling 1000 culture flasks (100 ml);
2. A unit process would involve a 100-liter fermenter.

Cultures of 1-5 liters are suitable for simulating commercial culture processes in a laboratory because at this scale of operation some parameters that can be ignored in low-volume cultures now become important and need to be controlled. These parameters include the method of oxygen supply, temperature control, pH control and culture mixing. A standard set of electrodes can monitor oxygen, pH and temperature. To obtain high cell densities and maximum productivity of secreted proteins, the type and design of the bioreactor and mode of operation need to be considered carefully.

Stirred Tank Reactor (STR)

A bioreactor is a term often used instead of a fermenter to describe equipment designed to grow cells in culture. The stirred tank bioreactor (or reactor) is the simplest and most widely used of all fermenter designs and consists of a cylindrical vessel with a stirrer (a pot and paddle). The design has been used extensively in all microbial fermentation and has been the main system used in yeast fermentation in the brewing industry for centuries. However, animal cells are more fragile and grow more slowly than most bacteria or fungi. They require gentler culture conditions and control systems that are optimized for lower metabolic rates.

Therefore, the design, mode of operation and control systems of an STR used for animal cells are distinctly different from those that would be applicable to bacterial or fungal cells. This is an important point because attempts at *ad hoc* modifications of a fermenter designed

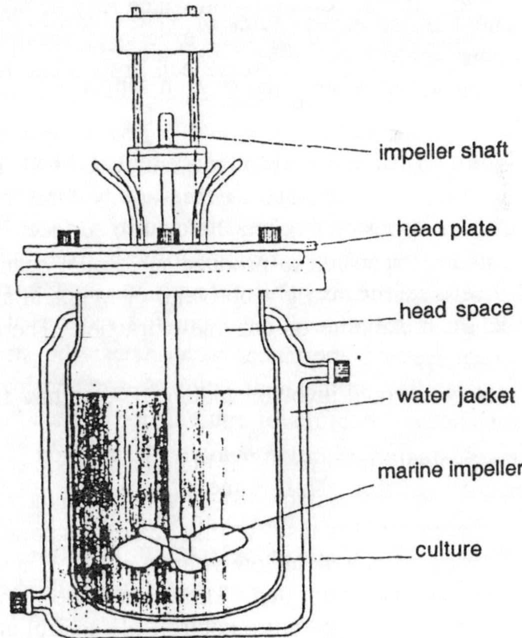

Fig. 5.1. Stirred tank bioreactor.

for bacteria are unlikely to be successful in growing animal cells. The STR has been developed commercially in large-scale animal cell culture processes up to a volume of at least 10,000 liters. For laboratory use there are also numerous bench-top STRs (1-5 liters) that are available commercially and that have been designed specifically for animal cell cultures. Small vessels (<20 liters) are made of glass whereas stainless steel is used for larger ones (>20 liters). Bioreactors like this one with a working volume of about 5 liters can be sterilized by autoclaving with the electrodes in position.

Bench-top models are generally made of glass with a stainless steel head plate, whereas the larger fermenters are made entirely from stainless steel. The metal head plate of an STR consists of a range of ports and pipes. This allows electrodes to be inserted and tubing to be attached for media input or sampling. Manufacturers of bench-top models include the following: Applikon, New Brunswick, LH Fermentation, Setric SGI, Braun, Bio-engineering. Each of these companies produces uniquely designed and controlled fermenters, all of which have been shown to be suitable for animal cell culture. The vessel (5 liter) at this scale can be made of glass whereas at higher volumes stainless steel is used as in the case of the 500-liter stirred bioreartor. There are several features of STRs that can be utilized to ensure adequate control of culture parameters such as agitation, temperature control, pH control and oxygen supply.

Agitation

The fragility of animal cells in culture has been a subject of considerable interest and debate in relation to fermenter design. Although cells in suspension can be damaged by various forces acting in a stirred culture, the major damaging force is from bubble bunting on the culture surface resulting from culture aeration. The hydrodynamic shear force resulting from the motion of a stirrer is thought to be of lesser importance; nevertheless the stirring speeds commonly adopted for animal cell cultures are considerably lower than those for bacterial cultures of equivalent volume.

The simplest stirring operation involves the rotation of a suspended bar by a magnetic stirrer. This is the system used in glass spinner bottles and is suitable for stiffing cultures up to a volume of 1 liter. At larger volumes, such magnetic stirrers are not suitable because of the increased energy required for rotation. Top-drive mechanical motors are normally used for STRs from bench-top models to the larger commercial fermenters. The stirring shaft fits through the stainless steel head plate and into the sterile culture. The stirring motor and outer part of the drive shaft can normally be disconnected from the head plate to allow the fermenter to be autoclaved. In early fermenter models, the stirring shaft was connected through the head, plate by replaceable rubber/silicone seals, which were vulnerable to damage and provided entry points of contamination. Later models have sealed units which are far more reliable.

Typically, maximum stirring rates of 100-150 rpm are used for cells in suspension, although this should be even lower (~40 rpm) for microcarrier cultures. Microcarriers are small spherical beads (diam. 200 μm) which can be suspended in a culture. Anchorage-dependent cells will attach and grow on the microcarrier surface.

In order to ensure adequate mixing at low stirring speeds, the culture vessels are designed with a round bottom, which distinguishes them from the flat-bottomed bacterial fermenters. Impeller blades which are fitted at the end of mechanical drive shafts are designed to allow vertical as well as horizontal liquid flow. The pitched-blade or marine type impellers are particularly suitable.

Temperature Control

High-power element heaters used in bacterial fermenters are avoided in mammalian cell cultures. At low agitation speeds the heat dissipation from such elements is insufficient to prevent localized heating in the culture. The temperature is usually controlled by a thermocirculator which pumps heated water (normally at 37°C) around an outer jacket or, for larger fermenters, through coiled pipes within the culture. An alternative system involves circulating warm air through an open-ended fermenter jacket. For low-volume bioreactors, external heating pads can be used to maintain the temperature.

pH Control

The optimal pH for animal cell cultures is around pH 7.4 and maximum cell growth occurs if this is maintained. However, without buffering or external control this pH could fluctuate during culture and may decrease to a level that would eventually inhibit growth.

An enriched CO_2 atmosphere as provided by an incubator can decrease these fluctuations. Here, pH is controlled by gassing the cultures with CO_2. However, this buffering capacity is limited and would be inadequate for the high cell densities that can normally be expected in wellcontrolled fermenters. Further pH control by gaseous CO_2 is limited by the available head space in some industrial bioreactors. Typically a 1-liter culture at 2×10^6 cells/ml requires a gas flow of 100 ml/min into a head space with a 1-liter capacity.

An alternative method of pH control is by direct acid or alkali addition from a peristaltic pump. Normally during growth only alkali addition is required to offset the net acidic production from cellular metabolism. Low-volume addition of a concentrated sodium bicarbonate is often used in preference to sodium hydroxide because of the danger of over-shooting the set-point with the stronger alkali. Thesc additions are normally governed by a computer-controlled pump or gas valve to a pre-set pH value. Acid (e.g. HCI) may also be added from an external source. However, this is usually not required as lactic acid is normally produced by cell metabolism and causes the culture pH to decrease during growth. The rotameters indicate the rate of gas flow into the culture, which can be controlled by the flow regulators. The electronic pH controller regulates the on/off function of the control valve.

Oxygen Supply

The supply of oxygen to satisfy cell metabolism is one of the major problems associated with culture scale-up. The oxygen consumption rate of mammalian cells varies from 0.06-0.6 mmol/liter/hour for a culture at 10^6 cells/ml. In small cultures (T flasks, etc.), this oxygen demand can be satisfied by gas diffusion from the head space through the culture surface. However, with increasing culture volume, the surface-volume ratio decreases. At cultures of 1 liter and above, the surface-volume ratio is generally too low to satisfy the overall oxygen demand at this cell concentration.

The surface-volume ratio of a bioreactor is normally defined by the aspect ratio. For a cylindrical-shaped bioreactor, the aspect ratio = diameter/ height of culture. If the cells utilize oxygen faster than it can be supplied, then the dissolved oxygen concentration of the culture media decreases eventually to a level that does not support cell growth. Thus, to avoid oxygen depletion, the *oxygen transfer rate* (OTR) across the liquid surface must be increased above the *oxygen utilization rate* (OUR) of the cells. In a simple fermenter containing 10^6 cells/ml and with an aspect ratio of one, oxygen supplied by diffusion will become limiting at around 1 liter. This is the critical volume at which OTR = OUR. To satisfy the OUR at higher culture volumes, a continuous supply of oxygen is required. The following points must also be considered in developing a system for oxygen supply:

1. The solubility of oxygen in an air-saturated aqueous solution is low (0.22 mM at 37°C). This is the maximum concentration of oxygen in culture media which is in equilibrium with air and is referred to as 100% air saturation;
2. Growth of many animal cell lines has been found to be optimal at a dissolved oxygen level below the maximum oxygen solubility and corresponding to 20-50% of air saturation.

Control of the oxygen supply

The dissolved oxygen level of a culture can be monitored by a sterilizable oxygen probe. This is fitted with a gas-permeable membrane to allow oxygen into an electrolyte contained in an inner chamber. The resulting reaction produces an electrical response that can be measured from a calibration based on differences between nitrogen-gassed (0%) and air-saturated water (100%). The dissolved oxygen can be controlled by the opening of a valve when the value decreases below a predetermined set-point. The action may increase the air flow in an airlift bioreactor (a), increase sparged oxygen in a stirred tank bioreactor (b), control the relative distribution of a combination of gases flowing into an airlift bioreactor (c) or sparge into a cell-free protected space of the bioreactor (d).

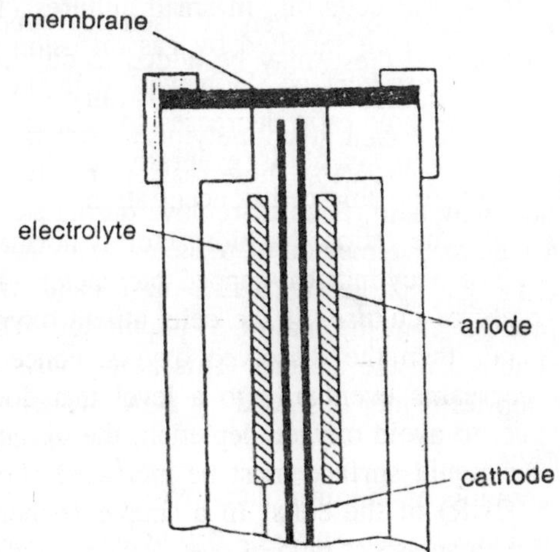

Fig. 5.2. Membrane-covered oxygen electrode.

Oxygen vs air

The OTR can be increased by filling the head space with oxygen instead of air. This increases the OTR by a factor of about five and increases the critical volume or cell density at which OUR = OTR. Air contains oxygen at ~21%. Therefore, the partial pressure of oxygen in air is 1/5th of that in pure oxygen. Oxygen solubility in culture media can also be increased (up to × 20) by the addition of some liquid perfluorocarbons. However, both these options are expensive and do not allow indefinite scale-up possibilities.

Sparging

Direct aeration by sparging is commonly performed in bacterial cultures and is by far the simplest operation for supplying oxygen. However, cell damage can be caused by this process. This damage is associated with gas bubbles bursting at the culture surface. Cells may attach to the bubbles as they are formed in the sparger and can be carried to the liquid surface where bubble bursting can cause a significant shear force to damage the cells. A further problem is that sparging through the high protein content of mammalian culture media can cause excessive foaming, particularly in serum-supplemented media. Serum-free media with a low protein content are less susceptible to foaming, although the cells are less well protected against damage associated with agitation. Intermittent (rather than continuous) sparging can reduce damage to cells.

Chemical protective agents can be used to reduce cell damage and foam formation and these may be added to cultures that may be susceptible to such problems. These are mainly synthetic polymers and include Pluronic (a mixed glycol polymer), polyvinyl alcohol (PVA) and polyvinyl pyrrolidone (PVP). These reagents can be useful in culture but some of them may be difficult to remove during subsequent product purification. An alternative method of reducing foam and bubble damage is to surround the gas sparger by a fine-mesh cage which can isolate the aeration process from the main cell population. This prevents the formation of bubbles in the culture. This is the basis of the Celligen unit—a bench-top fermenter designed by New Brunswick.

Indirect aeration

Indirect aeration of cultures can involve medium sparging in a secondary vessel. Such external oxygenation can alleviate the problems of air bubble or foam damage to cells and is particularly suitable for bioreactor configurations that involve media recirculation.

Diffusion via tubing

Oxygen can be supplied by gas diffusion through thin-walled silicone tubing. The tubing is submerged in the liquid medium and is supplied

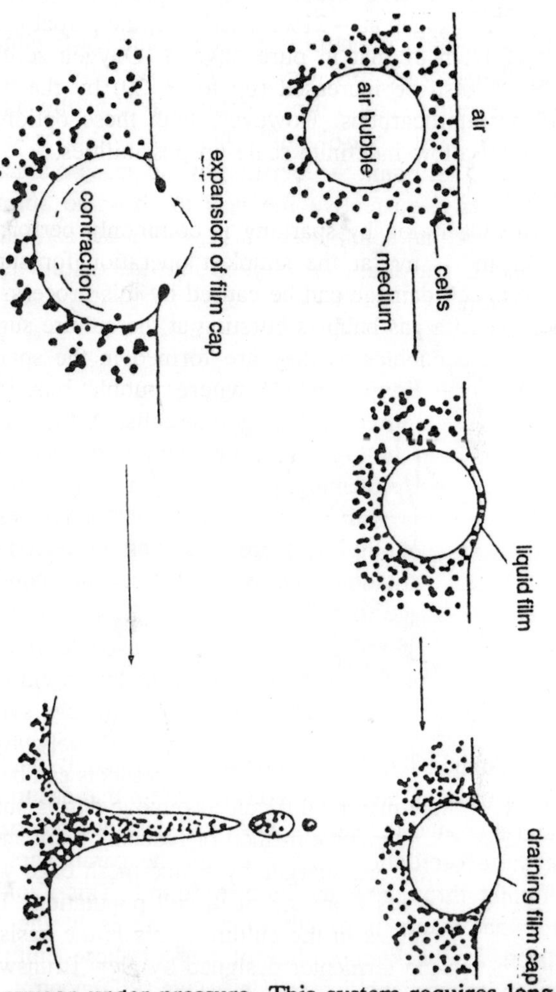

Fig. 5.3. Bubble bursting causes cell death on the liquid surface.

with air or oxygen under pressure. This system requires long lengths of tubing—at least 1 meter per 2 liters of culture. This may be unwieldy but can be overcome by adequate twining or binding of the tubing. Such a system, is available in fermenters designed by Applikon and Braun.

Process Control

Maximum productivity and consistency between runs can be maintained if critical culture parameters are controlled and maintained at their optimal values. This is helped by using sterilizable probes connected to computer control systems. A number of culture parameters, including temperature, pH, oxygen, and foam level, can be controlled by this method. The output from each probe is transmitted electronically to a control unit where the value is compared to a predetermined set-point. A deviation of the probe output from the set-point activates a corrective measure to shift the culture parameter towards the optimal conditions. This involves the activation of a pump to feed alkali, anti-foam, or oxygen into the culture or switch on a heating system.

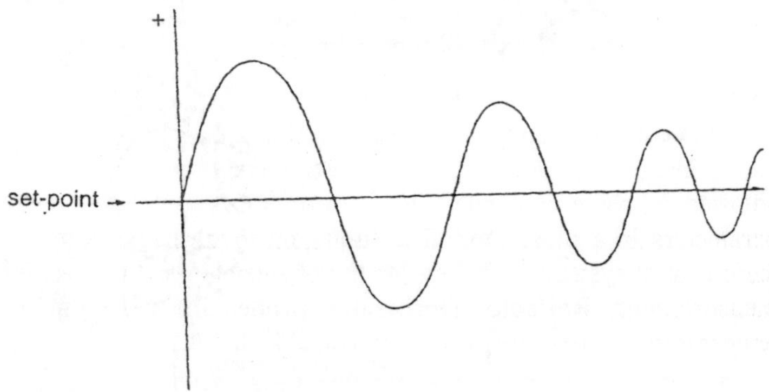

Fig. 5.4. Modulated feedback control.

A modulated feedback controller is commonly used in fermenter systems. This is an electronic control system that ensures against large fluctuations around the set-point. A well-designed feedback system can reduce these oscillations and maintain the parameters very close to the set-point throughout the culture period. This avoids the type of changes that occur in small-volume cultures in the laboratory. A modulated feedback controller of any of these culture parameters operates by proportional/integral/derivative (PID) control. This has three features of modulation which may be adjusted by proportional control, integral control or derivative time.

Proportional Control

The output of the controller is proportional to the error signal.

$$\theta = \theta_0 + k.E$$

where θ = output signal of the controller, θ_0 = output signal when the error is zero, k = controller gain or proportional band, and E = error or deviation from the set-point.

Integral Control

The output of the controller is a function of the integral of error and time. Here, the control action increases with time as long as the error is registered.

$$\theta = \theta + \frac{1}{t_1} \cdot \int E.dt$$

where t_1 = integral time constant.

Derivative Time

The output of the controller is a function of the rate of change of the error.

$$\theta = \theta_0 + t_d \cdot \frac{dE}{dt}$$

where t_d = derivative time constant.

PID combines these three features. The values of the proportional band, integral time constant and derivative time constant may be adjusted on some controllers so as to optimize control of the culture parameters in a bioreactor. The limitation to further development of these control systems is the availability of suitable probes for continuous measurement. Reliable, sterilizable probes for pH, oxygen and temperature control are presently available for cell cultures.

It would undoubtedly be advantageous to control other critical factors such as nutrient concentrations by feeding from pumps. This would allow nutrients such as glucose and amino acids to be maintained at their optimal concentrations in culture. Such an optimal environment for cell growth and metabolism could limit the accumulation of adverse metabolic products such as lactic acid and ammonia. pH decreases with lactic acid formation. Ammonia inhibits cell growth. Glucose and biomass probes are available but up to now have not been considered sufficiently reliable or cost-effective for continuous, routine use in animal cell fermenters.

Alternative Types of Bioreactors

Although the STR is the most widely used type of bioreactor for large-scale cell cultures, and certainly the type recommended for laboratory work (up to 5 liters), there are a number of alternative systems that have been developed for culture scale-up.

Airlift Fermenter

This type of fermenter consists of a tall column with an inner draught tube. Fluid circulation is provided by a stream of air which

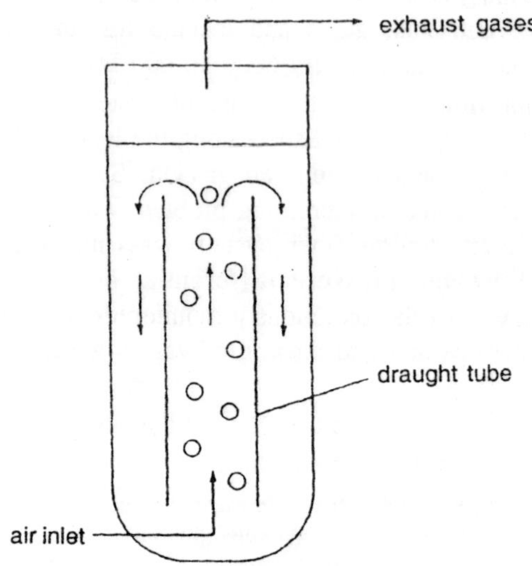

Fig. 5.5. Airlift fermenter.

passes through the inside of the draught tube. This is a simple system without mechanical components and therefore not susceptible to breakdown. Bubble or foam damage is minimized by having a long column, since it has been shown that maximum cell damage occurs at the point of bubble bursting at the top of the liquid column. An airlift fermenter ($>$1000 liters) has been used routinely for the production of bulk quantities of monoclonal antibodies from hybridomas by Lonza/ Celltech in the UK.

Hollow-fiber Bioreactor

This consists of bundles of synthetic, semipermeable hollow fibers, which offer a matrix for cell growth similar to the vascular system *in vivo*. The bioreactor is based on equipment originally developed for hemodialysis with a cartridge containing several thousand minute capillarylike plastic tubules (fibers) with perfusable membrane walls. Liquid can flow through the fibers (the intracapillary space) or through the space between the fibers (the extracapillary space). In the normal operation culture medium is pumped through the intracapillary space and a hydrostatic pressure permits the exchange of nutrients and waste products across the capillary wall. The cells and large-molecular-weight products are held in the extracapillary space. They grow to very high densities equivalent to the characteristics of tissue (up to 10^9 cells/ml) as could be seen from micrographs of across-section of several fibers. Medium is pumped into the fibers creating a relatively high pressure within the intracapillary space forcing nutrients through the pores into the extracapillary space. The perfusion of the medium provides efficient exchange of nutrients, gases and metabolites. the molecular weight cut-off of the fiber is normally 6000-10,000 Da.

A major limitation of this type of system is that the pressure difference that may be established along the length of fibers can cause nutrient gradients and uneven cell growth. Such pressure differences and gradients become an increasing problem with scale-up. The design of the 'Acusyst' hollow-fiber system was intended to correct this problem. Here the pressure differential between the intro- and extracapillary space is continuously monitored by sensors which serve to control the opening and closing of valves which in turn affect the capillary pressure.

The exchange of molecules through the fiber wall occurs in phases governed by a cyclic mode of pressure changes which prevent undue gradients developing along the fibers. This hollow-fiber system is suitable for both anchorage-dependent and -independent cells. Continuous operation allows a high rate of product recovery from a stationary high-density culture held in the extracapillory space over a long period of time. Hollow-fiber bioreactors can be operated continuously for several months.

Packed-bed or Fixed-bed Bioreactor

This type of bioreactor consists of a static support matrix for the attachment and growth of anchorage-dependent cells. Cell growth is supported by a continuous flow of medium. The design of these bioreoctors attempts to maximize the surface area available for growth in a given volume. Three examples of bioreactors designed for such a purpose are given.

Glass bead column

The glass beads offer a good surface for cell attachment. Cells are allowed to attach and grow on the bead surface. The medium is recirculated through the packed bed by a pump and oxygenated by an input of air in a secondary vessel. An attempt is made to ensure on even distribution of cells over the matrix so as to maximize yield and to prevent any concentration gradient of cells in the system. In the column of glass beads this problem may be prevented by inoculating cells through a perforated tube, placed centrally through the vertical height of the column. Precoating the glass beads with fibronectin can ensure rapid cell attachment.

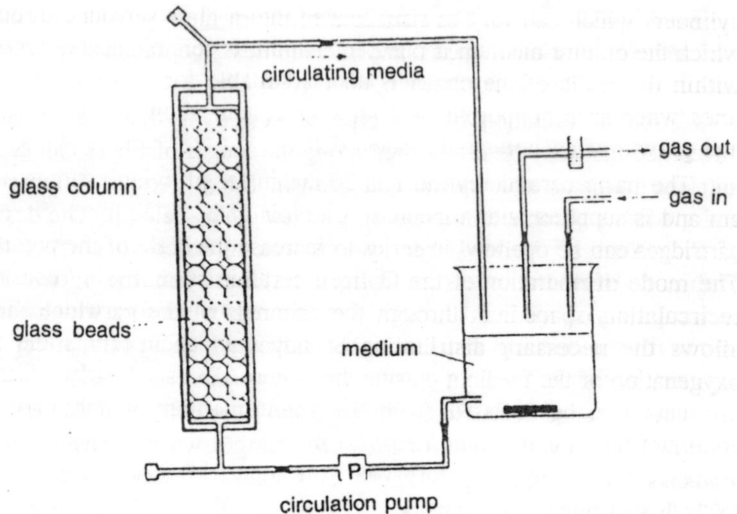

Fig. 5.6. Packed-bed bioreactor.

Ceramic bioreactor

Although now largely in disuse the commercially developed 'Opticell' system was an attempt to offer a large surface area in a series of multiple channels, which run through the length of a ceramic cylinder. Each channel is a square with 1 mm sides and an inner

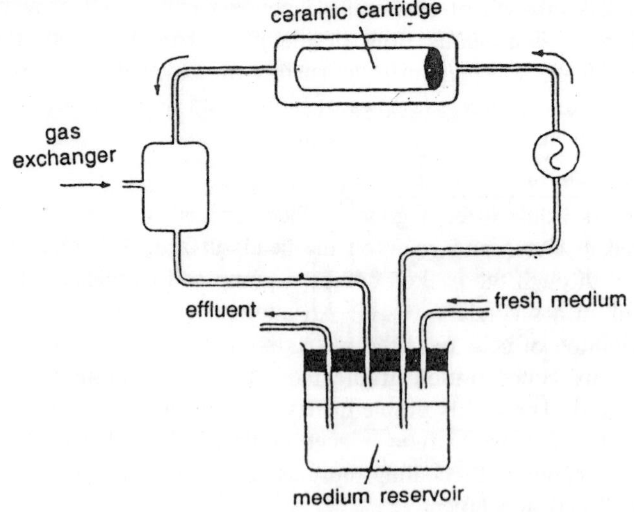

Fig. 5.7. Opticell ceramic system.

surface area for cell attachment. A cartridge consists of a ceramic cylinder, which can vary in size, loaded into a glass surround through which the culture medium is pumped. A porous ceramic allows growth within the walls of the channels and is suitable for nonadherent cell lines whereas a nonporous smooth surface can allow the attachment and growth of anchorage-dependent cells.

The basic ceramic cylinder is 30 cm in length with a radius of 4 cm and is supplied with a medium reservoir of up to 20 liters. Several cartridges can be operated in series to increase the scale of the process. The mode of operation of the Opticell ceramic system is by constant recirculation of medium through the channels of the cartridge. This allows the necessary distribution of nutrients and permits the re-oxygenation of the medium outside the culture. Extracellularly released products can be isolated from the spent medium which may be continuously siphoned into a harvest flask from the main reservoir at a preset rate. The main reservoir is replenished by fresh medium from a feed tank.

Cell cube

This consists of a stack of 20 cm^2 polystyrene plates spaced 1 mm apart by rigid spacers and in a case that can be sterilized. The cells attach to either side of each polystyrene plate and are supported by the flow of culture medium between the plates. In the cell cube multiple inlet ports ensure the upward flow of liquid medium. Each port is shaped like a nozzle to ensure maximum distribution of the liquid over the plates. These design features ensure the even distribution of cells over the available surface. Such packed-bed or fixed-bed

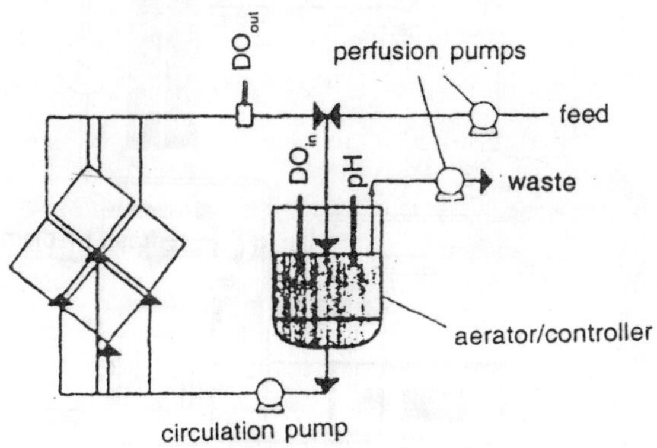

Fig. 5.8. Cell cube bioreactor.

bioreactors are suitable for scale-up and commercial production of vaccines from the viruses that can be propagated on the cells. A commonly used alternative for anchorage-dependent cell growth is the use of microcarriers which are small beads that can be suspended in culture and offer a surface for cell attachment. These are pseudo-suspension cultures and can be conducted in stirred tank bioreactors.

Fluidized-bed Reactor

In this system, cells immobilized or entrapped in beads are held in suspension in a column by an upward flow of liquid medium that is recirculated by pumping. Some fluidized-bed reactors have been designed in a funnel shape with a fluid flow from the narrow bottom. The density and bead size are generally high so that a linear fluid velocity supports the beads in a relatively homogeneous suspension. This can be compared to the use of microcarriers in a stirred tank reactor. The bead characteristics need to be optimized to suit the bioreactor. Porous collagen and alginate beads have been used successfully. Bead densities can be increased as required by the addition

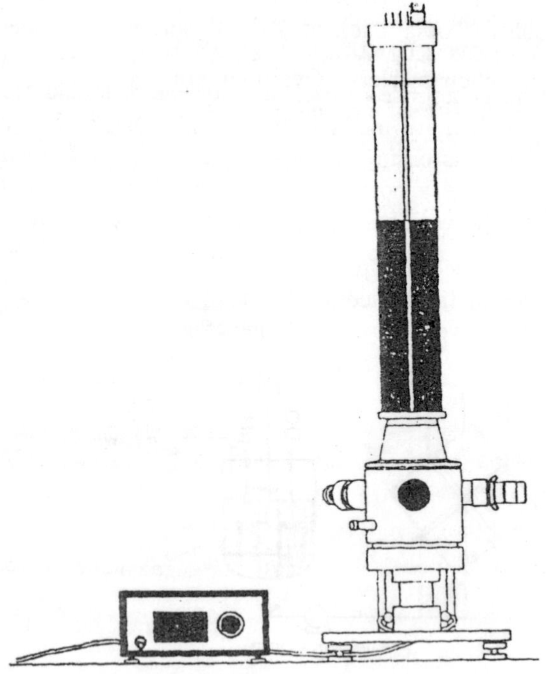

Fig. 5.9. Cytopilot mini.

of a heavy metal. In the example shown the fluid circulation is integrated into a single module.

The module consists of an upper and lower cylindrical chamber. The lower chamber consists of a magnetic stirrer, probes, a sample and heating jacket allowing the flow of temperature-controlled water. This is separated from the upper chamber by a porous distributor plate. The upper chamber is a tall column containing the porous microcarriers suspended in culture medium and into which the cells ate entrapped. Liquid is allowed to circulate via a cell-free inner tube by a magnetically driven stirrer, which controls the degree of fluidization of the microcarriers. The microcarriers used in this bioreactor are macroporous, allowing easy entry of the cells on inoculation and an inner capacity to allow growth to high cell densities. The microcarriers (Cytoline) are weighted to a density of around 1.3 g/cm^3 allowing high sedimentation rates in the fluidized column. The bioreactor is run in a perfusion mode, allowing a continuous flow of effluent containing the cell product. The culture is simultaneously provided with an equivalent flow of fresh medium to maintain a constant culture volume in the bioreactor.

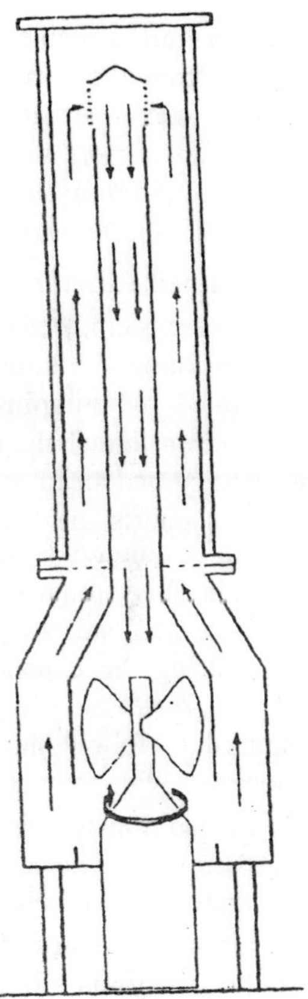

Fig. 5.10. Fluid flow within a fluidized bed bioreactor.

6

Monitoring Culture

Growth in cultures is normally determined by counting cells at regular intervals. In routine cultures this would be performed once a day, which corresponds to the approximate doubling time of mammalian cells during the exponential growth phase. This would establish an overall growth profile of a culture. More frequent counts would be required to follow more subtle changes that may, for example, be associated with the cell growth cycle.

In addition to this, it is worthwhile inspecting cultures under a microscope to determine if there are any significant changes to the appearance of the cells.

Growth in cultures is normally determined by counting cells at regular intervals—at least once a day. The two direct methods commonly used are manual counting through a microscope or electronic counting by a particle counter. Both methods depend upon obtaining a sample of an even distribution of cells in suspension. Therefore it is extremely important to ensure that the culture is well mixed by stirring or shaking before taking a sample.

Hemocytometer

This is a thick glass plate which fits onto the adjustable stage of a microscope. There are several designs available. The most common type used is the improved Neubauer. Through the microscope you can see a grooved calibrated grid on the hemocytometer surface. This consists of nine large squares with 1 mm sides. The central square is subdivided into 25 squares each of which is further subdivided into 16 squares of 0.05 × 0.05 mm. A cell suspension is put onto the grid by touching the end of a capillary tube containing the cell suspension at

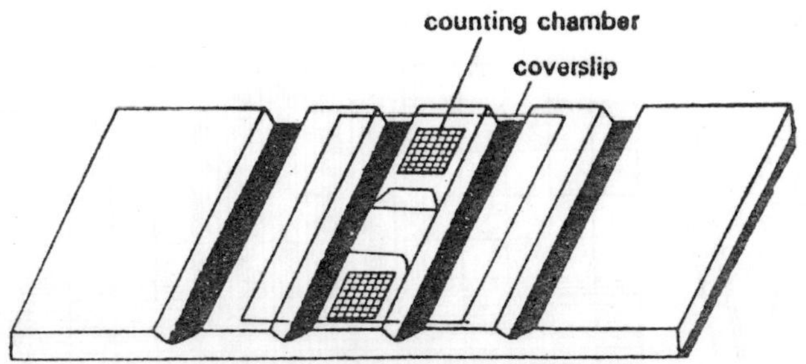

Fig. 6.1. Glass slide with counting chambers.

the edge of a cover slip placed on the upper surface of the hemocytometer. The position of the cover slip forms a chamber, the sides of which are the calibrated edges of the slide raised exactly 0.1 mm above the ruled surface. The volume contained in each of the nine large squares is $1 \times 1 \times 0.1 = 0.1$ mm^3 (= 0.1 μl). The cells are then counted in a standard volume (usually 5×0.1 μl) as defined by the area of the grid. In this case:

$$\text{Cells per ml} = \frac{\text{Total count} \times 10^4}{5}$$

A standard microscope at low magnification of $\times$ 40 to $\times$ 100 is required for viewing the cells. A hand-held tally counter helps in counting.

Trypan blue is often added to the cell suspension before counting. The dye penetrates the membrane of nonviable cells which are stained blue and which can therefore be distinguished from viable cells. Viability gives an indication of the ability of a cell to divide. The viability is usually expressed as a percentage of viable cells in a population.

% viability = (number of viable cells/total number of cells) $\times$ 100

This is the most commonly used assay for cell viability.

A modification of this method involves counting nuclei. Incubation of cell samples in a mixture of citric acid and crystal violet causes cells to lyse and the released nuclei to stain purple. Nuclei counting is well suited to the determination of anchorage-dependent cells, for example when attached to microcarriers. However, care must be taken in interpreting nuclei counts as cells can become binucleated, particularly, when growth is arrested. As a result the nuclei concentration may be higher than the cell concentration.

The hemocytometer counting method is simple and effective but can be laborious if many samples are being analyzed. The expected

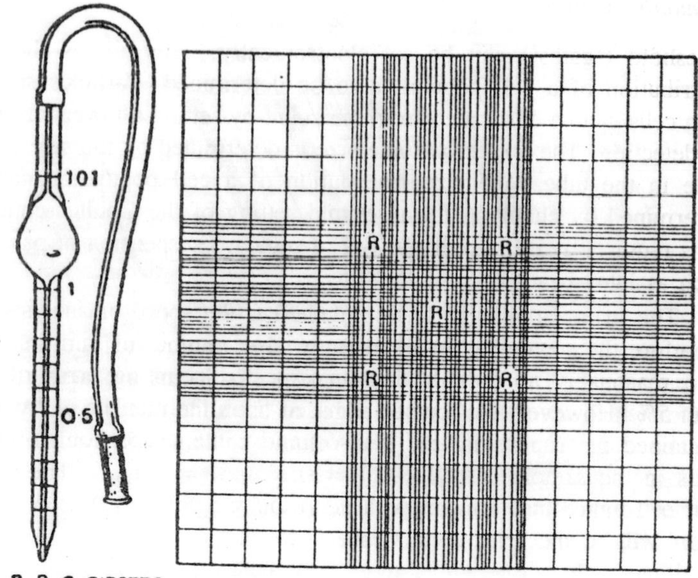

Fig. 6.2. R.B.C. pipette and small squares in the counting chamber. R denotes small squares for red cell count.

error in cell concentration is around 10%. This can arise from variability in sampling, dilution, mixing, filling the chamber or counting.

Coulter Counter

An electronic cell counter (or Coulter counter) is suitable for rapid counting of multiple samples of cells in suspension. The principle of the counter is that a predetermined volume (usually 0.5 ml) of a cell suspension diluted in buffered saline is forced through a small hole (diameter 70 μm) in a tube by suction. The cells (or any particles) cause a measurable change in electrical resistance as they pass between two electrodes, one inside and one outside the glass tube. This produces a series of pulses recorded as a signal on an oscilloscope. Several thousand particles per second may be counted and sized with accuracy.

In the simplest instrument (e.g. Coulter model D Industrial) a lower size threshold of counting may be set electronically. Thus, particles smaller than cells (dust or cell fragments) can be eliminated from the count. The largest particle size is determined by the size of the orifice in the tube, which is normally 75 or 100 μm. For routine counting the Coulter Industrial D is more than adequate. In more complex instruments (e.g. Coulter model ZB) a lower and upper size threshold ('gates') can be set electronically. This allows the size distribution of a cell population to be determined. Particles smaller than cells can be eliminated from the count by setting a lower threshold of detection. The largest particle size is determined by the size of the hole in the tube. The size distribution of a cell population can be determined by changing the threshold setting of the Coulter counter. This requires a size calibration with a standard suspension of particles such as latex beads or pollen grains.

The major advantage of this method is the speed of analysis and therefore the suitability for counting a large number of samples. It is also reasonably accurate within an expected counting error of less than 5%. However, the method is based upon the number of particles contained in suspension and, consequently, the proportion of viable cells in the sample cannot be determined. Also, it must be ensured that cell aggregates are not present in the sample; otherwise the cell count will be underestimated. Cell suspensions above 10^4 per ml should be diluted otherwise two or more cells passing through the counting zone at the same time will cause coincidence errors. A standard protocol is to dilute a cell suspension (×20-40) in a saline solution (Isoton).

Indirect Methods of Cell Determination

Indirect methods of estimating cell growth rely on the measurement of an intracellular cell component such as DNA or protein or alternatively an extracellular change such as nutrient depletion or an enzyme activity released by the cells. Indirect methods of growth estimation depend upon a correlation between the measured parameter and cell concentration. However, it is important to realize that these relationships are rarely linear over the course of a culture. It is well documented that the total protein content and specific enzyme activity levels measured on a per cell basis vary substantially over the course of a culture due to changes in the growth rate and composition of the culture medium. For example, the protein and enzyme content per cell will be high during exponential growth but lower in the lag of stationary phases.

In some situations, as may occur for example in immobilized cell bioreactor systems, an indirect measurement of cell growth may be the only option available. This can be used to monitor the progress of a culture. However, care must be taken if such data are used in comparative analysis between cultures, as differences may be a reflection of changes in metabolic or functional activity rather than of cell concentration. There are a number of colorimetric methods based on the measurement of cell components. These are relatively simple methods and suitable for multiple samples.

Protein Determination

Total cell protein can be used as a measure of biomass (total cellular material) but the protein content per cell can vary during cell culture and this can be a reflection of changing cell size. The protein content of a mammalian cell is typically 100-500 pg/cell. These measurements are also useful in the expression of specific enzyme activities which are commonly expressed as the maximum measured reaction velocity of an enzyme per total cell protein.

The most sensitive colorimetric assays are the Lowry and Bradford methods. Of these the Bradford assay is favoured because of speed, sensitivity and negligible interference from other cell components. By this method, lysed cells are solubilized in NaOH and added to the reagent, Coomassie blue. A blue colour which develops within 10 min can be measured by a colorimeter or spectrophotometer and compared with standard proteins.

DNA Determination

The DNA content of diploid cells is usually constant, although variations can occur as a result of the distribution of cells through the cell cycle. Cells in the G1 phase have the normal diploid content of DNA, which is typically 6 pg per cell. DNA measurement is probably the best method to use as an indicator of the number of cells in a solid tissue. A commonly used method involves treatment of the solubilized cells with diphenylamine reagent which forms a blue colouration with DNA. Fluorescence detection can also be used with Hoechst 33258 or DAN, which bind to DNA.

Glucose Determination

If cells are immobilized within a bioreactor, then it is often difficult to obtain a cell sample. Examples of such bioreactors include a packed-bed or hollow fiber system. Cell growth can be monitored by changes in the concentration of key components of the culture medium. The rate of change in the glucose content of the medium may be suitable for such an assay as an indirect measure of cell concentration. Alternatives include measurement of lactic acid production or oxygen consumption.

Correlations have been shown between cell concentration and consumption or production of these components. This relationship may be constant for a particular cell line under a given set of conditions. However, if the cell line or any of the culture conditions are altered the relationship between substrate consumption or product formation and cell number will change.

Viability Measurements

Viability is a measure of the metabolic state of a cell population, which is indicative of the potential of the cells for growth. This is a measure of the proportion of live, metabolically active cells in a culture, as indicated by the ability of cells to divide or to perform normal metabolism. The viability may be determined from simple assays such as dye exclusion where cells are designated as either viable or nonviable. From the dye exclusion method the loss of viability is recognized by membrane damage resulting in the penetration of the dye, trypan blue. Other dyes that can be used include erythrosin B, nigrosin and fluorescein diacetate.

Tetrazolium Assay

The tetrazolium assay is a measure of cellular oxidative metabolism. A colorimetric assay for viable cells has been developed using the tetrazolium dye, MTT [3-(4,5-dimethylthiazol-2-yl)-2,5-diphenyltetrazolium bromide]. The dye is cleaved to a coloured product by the activity of dehydrogenase enzymes and this indicates high levels of mitochondrial activity in cells. The colour development (yellow to blue) is proportional to the number of metabolically active cells. However, there is considerable variation in results between cell lines with some cells producing a very low response. Cells with a significant level of oxidative metabolism such as CHO produce a good response in the MTT assay. The method is particularly convenient for the rapid assay of replicate cell cultures in multi-well plates.

Lactate Dehydrogenase Determination

Loss of cell viability may be followed by an increase in enzyme activity in the culture medium caused by a damaged cell membrane which allows large molecules to enter or leave the intracellular pool. The activity of lactate dehydrogenase (LDH) is the most commonly measured enzyme in this technique. This activity increases as the enzyme escapes through the damaged membrane of nonviable cells. The enzyme activity can be measured easily by a simple spectrophotometric assay involving the oxidation of NADH in the presence of pyruvate. The reaction is monitored by a decrease in UV absorbance at 340 nm.

$$\text{pyruvate} + \text{NADH} + \text{H}^+ \xleftrightarrow{\text{LDH}} \text{lactate} + \text{NAD}^+$$

NADH absorbs at $\lambda = 340$ nm.

The specific enzyme activity is defined as the maximum measured reaction velocity of an enzyme per total cell protein. For LDH this equals $\sim 1\text{–}3 \times 10^3$ nmol lactate/min per mg cell protein. The method is well suited for the determination of multiple samples, particularly if a multi-well plate reader is available. This instrument enables the light absorbance of all wells to be measured within seconds. However, care must be taken when interpreting the results by this method because the LDH content per cell can change considerably during the course of batch culture. The loss of cell viability can be expressed as the activity of LDH in the medium as a proportion of total LDH in the culture.

Adenylate Energy Charge

The energy charge is an index based on the measurement of the intracellular levels of the nucleotides, AMP, ADP, and ATP.

$$\text{the energy charge} = \frac{\text{ATP} + 0.5.\text{ADP}}{\text{ATP} + \text{ADP} + \text{AMP}}$$

This is based on the interconversion of the three adenylate nucleotides in the cell:

$$\text{AMP} \leftrightarrow \text{ADP} \leftrightarrow \text{ATP}$$

This index varies between the theoretical limits of 0 and 1. For normal cells, values of 0.7-0.9 would be expected, but a gradual decrease in the value gives an early indication of loss of viability by a cell population. These nucleotide concentrations can be measured by high performance liquid chromatography (HPLC) or by luminescence using the luciferin-luciferase enzyme system. Such measurements are not as easy to perform as the routine counting procedures discussed earlier but can be a means of monitoring the decline in the energy metabolism of a cell culture during the loss of viability.

Rate of Protein or Nucleic Acid Synthesis

The metabolic activity of cells can be measured by their rate of protein or DNA synthesis. This is based on the incubation of intact cells in standard culture medium to which is added a radioactively labeled amino acid or a nucleotide. Any radioactive amino acid is suitable but those most commonly used are ^{3}H-leucine or ^{35}S-methionine. Nucleic acid synthesis is measured by the utilization of tritiated thymidine (^{3}H-TdR). This is also the basis for the determination of the cell cycle.

Colony-forming Assay

The most precise of all methods of viability measurement is the colony-forming assay. Here the ability of cells to grow is measured directly. A known number of cells at low density is allowed to attach and grow on the surface of a Petri dish. If the cell density is kept low, each viable cell will divide and give rise to a colony or cluster of cells. From this, the 'plating efficiency' is determined as the number of colonies scored per 100 cells plated × 100. Although the colony-forming assay is time-consuming, it has been used widely in cytotoxicity studies. The effect of treating cells with a potentially toxic compound is determined by a change in the plating efficiency. The colonies of treated cells are compared with controls.

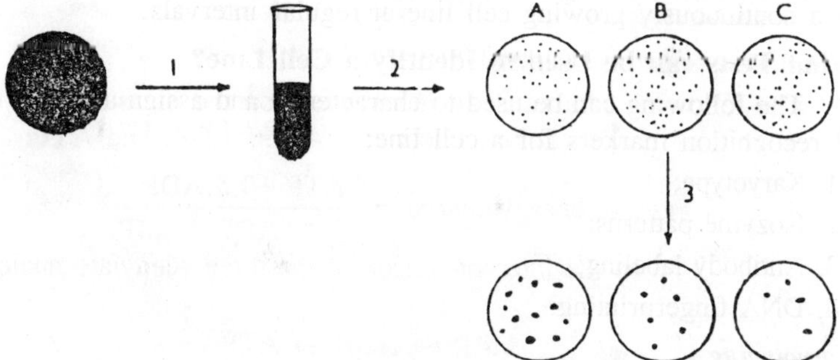

Fig. 6.3. Colony-forming assay. 1. Remove the cells from the attachment surface and suspend in media. 2. Add cell suspension to Petri dishes, each may be treated differently. A is the untreated control B and C are treated with agent to tested. Incubate to allow colonies to form. 3—Fix the colonies with methanol/acetic acid and stain with crystal violet. The relative surviving fraction is determined from the number of independent colonies in each dish compared to the control.

The effect of the treatment is assessed by the relative surviving fraction which equals the colonies on treated Petri dish/colonies on untreated Petri dish. A less precise method of determining the viability by the cellular reproductive potential is from the lag phase of a growth curve. By extrapolation from the linear portion of the growth curve to time zero, the derived cell number can be compared with the original count. This method can be easily adapted to determine how a particular treatment affects cell viability.

Cell Line Identification

Need for Cell Line Monitoring

Because different cell lines may be handled in the same laboratory, there is a danger of cross-contamination. This problem became apparent in the 1960s when it was discovered that a variety of widely used cell lines that had been exchanged between laboratories were contaminated with HeLa cells. HeLa are fast-growing cells that are capable of outgrowing most other cell lines. This contamination was so widespread that many cell lines that were originally isolated in the 1960s have been shown to have genetic markers derived from HeLa cells, indicating that DNA exchange occurred at some time.

Cell lines that are grown continuously in culture are vulnerable to genetic change. This can arise through intrinsic alteration of the genome of the cultured cells by mutation or by selective pressures. Therefore, cells grown for several passages in culture may be quite different from those originally derived from primary tissue. Because of these reasons, it is important to monitor the genetic characteristics of a continuously growing cell line at regular intervals.

What Tests can be Used to Identify a Cell Line?

The following can be used to characterize and assign a unique set of recognition markers for a cell line:

1. Karyotype;
2. Isozyme patterns;
3. Antibody labeling;
4. DNA fingerprinting.

Karyotyping

This is used to establish the distribution of chromosomes (*karyotype*) and may indicate the species of origin of the cells. The chromosome composition can also Indicate whether the cells are transformed or whether any chromosome damage has occurred. Individual chromosomes can be seen under a microscope during mitosis. Chromosome identification can be aided by a banding technique. This involves the use of a dye which generates a reproducible pattern of stained bands on each chromosome. The molecular basis for the development of these bands is not clearly understood but probably relates to areas of highly condensed chromosomal material. One commonly used dye is Giemsa. In this technique nucleoprotein martially digested by trypsin, and the Giemsa dye produces a characteristic pattern of G

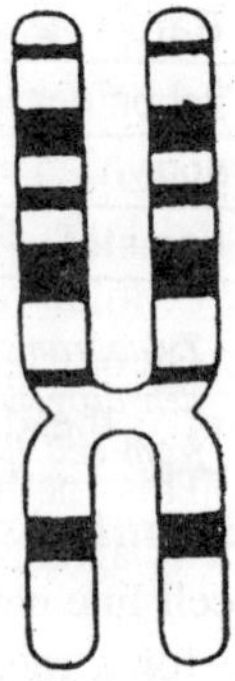

Fig. 6.4. G-band of a human X chromosome.

bands. A transformed cell population may be heteroploid, that is, have a chromosome count that varies between individual cells. In order to characterize such a cell line, chromosomes are counted for 100-200 individual cells, and a modal chromosome number is calculated.

Isozymes

The analysis of isozyme patterns of selected enzymes has been used extensively as a technique to identify the species of origin of cell lines with a high degree of certainty. Isozymes (or ***isoenzymes***) are structurally different forms of the same enzyme. They catalyze the same reaction but have different protein structures. The technique involves gel electrophoresis of cell homogenates under nondenaturing conditions. Specific activity stains are used to develop a banding pattern of isozymes (*zymogram*), which is characteristic of a particular cell line. Such zymograms are usually photographed to provide a permanent record of each cell line. Of the various enzymes that can be separated into isozymes by this technique, glucose 6-phosphate dehydrogenase, lactate dehydrogenase, and nucleoside phosphorylase have been particularly well characterized. By using several enzymes the distinguishing features of a cell line are established. These features can often distinguish cell lines even if derived from the same species. This is a more rapid technique compared to karyotyping and requires smaller cell samples.

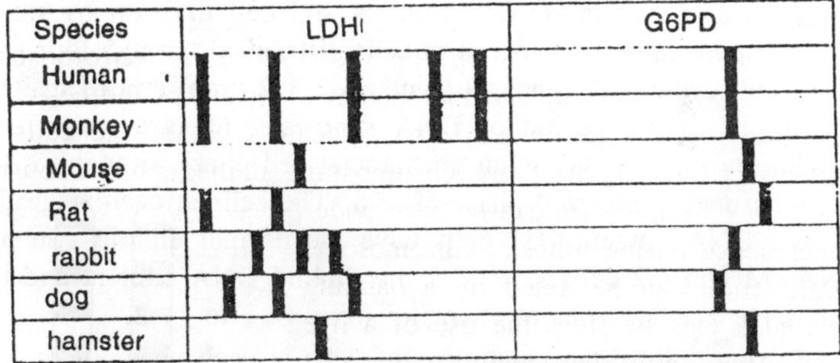

Fig. 6.5. Zymograms. Distinct banding patterns are shown for the enzymes lactate dehydrogenase (LDH) and glucose 6-phosphate dehydrogenase (G6PD) derived from cell lines of different species.

Labeled antibodies

A cell line can be identified by use of a fluorescent-labeled antibody specific for a membrane antigen. The antibody, prepared by methods is conjugated to a suitable fluorescent compound, suck as fluorescein. The conjugate will bind specifically to the outer membrane of the chosen cells which can be identified by fluorescence microscopy or by a fluorescence-activated cell sorter.

DNA fingerprinting

This technique is well known in forensic science but is gradually being adopted as a standard reference technique for cell line identity in culture collections. A unique DNA 'fingerprint' can be developed for a particular cell line. The DNA fingerprint results from the fragmentation pattern produced by digestion of cellular DNA with restriction endonucleases. The resulting restriction fragment digest is separated by electrophoresis. Radioactive probes are then used to hybridize to specific restriction fragments which can be highlighted by autoradiography. This results in a characteristic 'bar-code' pattern.

The most useful probes for this purpose are those that hybridize to 'mini-satellite' DNA. These are repetitive nucleotide sequences of varying length found throughout the genomic DNA. Certain restriction enzymes are used because they are known to cut DNA within the minisatellite regions. The length and distribution of the resulting minisatellite DNA fragments are unique to individuals and hence can be used for identification, including cell line identification.

Analysis of the Cell Cycle

During growth, cells go through a series of stages characterized by two major events—DNA synthesis and cell division by mitosis. Any cell population will consist of cells distributed through four phases of the cell cycle—G1 (gap), S (synthesis), G2 (gap) and M (mitosis). The S phase is a period of DNA synthesis, M is a stage of cell division by mitosis, G1 is an uncharacterized phase after mitosis and G2 is an uncharacterized phase after S. The cells at each stage differ in their-DNA content—G1 cells have the normal diploid content of DNA. M and G2 cells have $\times$ 2 the diploid DNA content and the S cells have a DNA content changing from $\times$ 1 to $\times$ 2.

During culture the proportion of cells at each phase of the cycle changes. In the stationary'and decline culture phases there is an increased proportion of cells at G1 of the cell cycle. During growth there is a high proportion of cells in the S phase and a low proportion in the G1 phase. The metabolic activity of cells can often be related to cell cycle phase changes. For example, enhanced antibody synthesis has been shown by some lymphocyte hybridomas in the G1 phase. The following two methods are available to analyze the distribution of a cell population between the phases of the cell cycle.

Tritiated Thymidine Pulse Method

This method can be used to determine the proportion of cells involved in DNA replication and allows an estimate of the length of the S phase. Cells are incubated in a medium containing ^3H-thymidine for a short period of time (~30 min). During this time only cells in the S phase will incorporate radioactivity into the DNA. This is followed by a period (~ 1 h) of incubation in nonradioactive medium. The cells are then fixed on a microscope slide and prepared for autoradiography. This allows the determination of the fraction of cells which have incorporated ^3H-thymidine. This fraction multiplied by the total generation time gives the length of the S phase.

Flow Cytometer

This is a sophisticated and expensive piece of equipment which would be found in a specialized laboratory. The cytometer is capable of analyzing cells treated with a fluorescent stain—for cell cycle analysis the stain is specific for DNA. The stained cells are then forced through a fine nozzle at a rate of several thousand cells per

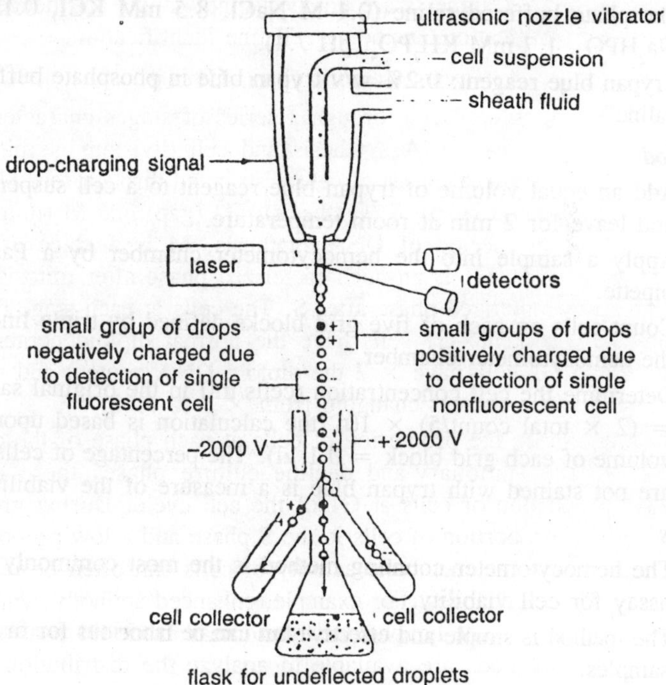

Fig. 6.6. Fluorescence-activated cell sorter.

second. This stream of cells is exposed to a laser beam and the resulting fluorescence emission is detected by a photomultiplier. Since the fluorescence intensity is directly proportional to the DNA content of each cell, the data output gives a distribution analysis of the cell population by DNA content. This information can be related to the distribution of cells through the cell cycle. The G2 + M peak is recognized at ×2 the relative DNA content of the G1 peak, whilst the S phase cells are indicated by the band between the peaks. A further application of this type of instrument is in fluorescent-activated cell sorting. Cells detected with a fluorescent label may be differentially charged and that allows them to be deflected to a collecting tube following passage through two electrodes.

PROTOCOL

Determination of the Concentration of Viable Cells in a Suspension by Hemocytometer Counting

Materials

Phosphate buffered saline (0.1 M NaCl, 8.5 mM KCI, 0.13 M Na_2HPO_4, 1.7 mM KH_2PO_4, pH 7.4).

Trypan blue reagent: 0.2% w/v trypan blue in phosphate buffered saline.

Method

1. Add an equal volume of trypan blue reagent to a cell suspension and leave for 2 min at room temperature.
2. Apply a sample into the hemocytometer chamber by a Pasteur pipette.
3. Count cells on each of five grid blocks defined by triple lines in the hemocytometer chamber.
4. Determine the cell concentration (cells/ml) in the original sample = (2 × total count/5) × 10^4 (the calculation is based upon the volume of each grid block = 0.1 μl). The percentage of cells that are not stained with trypan blue is a measure of the viability.

Notes

1. The hemocytometer counting method is the most commonly used assay for cell viability.
2. The method is simple and effective but can be laborious for multiple samples.
3. At least 100 cells should be counted for statistical validity of the final value.

Determination of the Nuclei Concentration in a Culture Sample

Materials

Crystal violet reagent:0.1% w/v crystal violet in 0.1 M citric acid.

Method

1. Allow microcarriers from a culture sample (1 ml) to settle to the bottom of a centrifuge tube.
2. Remove clear supernatant by aspiration.
3. Add 1 ml of crystal violet reagent.
4. Incubate at 37°C for at least 1 h.
5. Introduce a sample into the hemocytometer chamber and count the purplestained nuclei as for whole cells.

Notes

This method can be used to determine the nuclei concentration of anchoragedependent cells attached to microcarriers. Care must be taken in interpreting nuclei counts as cells can become binucleated, particularly when growth is arrested. As a result the nuclei concentration may be higher than the cell concentration.

Determination of Cell Concentration by a Coulter Counter

Materials

Saline solution –0.7% NaCl, 1.05% citric acid, 0.1% mercuric chloride in distilled water.

Method

1. Add 0.5 ml of a cell suspension (10^5-10^4 cells/ml) to 19.5 ml of the saline solution.
2. Introduce the suspension into a Coulter counter.
3. From standard settings of the counter 0.5 ml of the suspension is counted. Multiply this count by 40 to give the original cell concentration.

Notes

1. The major advantage of the Coulter counter method is the speed of analysis and it is therefore suitable for counting a large number of samples.
2. The method is based upon the number of particles contained in suspension and, consequently, the proportion of viable cells in the sample cannot be determined.
3. It must be ensured that cell aggregates are not present in the sample; otherwise the cell count will be underestimated.

4. The Coulter counter can also be used to determine the size distribution of a cell population by careful control of the threshold settings of the instrument.

Determination of the Protein Concentration of a Cell Suspension

Materials

Bradford's reagent: Dissolve 100 mg of Coomassie brilliant blue G in 95% ethanol (50 ml) and 85% phosphoric acid (100 ml). After the dye dissolves make the solution up to 1 liter with distilled water. Alternatively, a dye (Coomassie) reagent liquid concentrate can be purchased from Biorad.

Method

1. Homogenize or sonicate a cell suspension (10^6 cells/ml).
2. Add 5 ml Bradford's reagent to 100 μl of the lysed cell sample (0-0.5 mg/ml protein).
3. Incubate for 10 min at room temperature.
4. Measure the absorbance at 595 nm.
5. Determine the sample concentrations from a standard curve, which is established from standard solutions of bovine serum albumin (BSA) at 0-0.5 mg/ml protein.

Determination of the DNA Concentration of a Cell Suspension

(a) Hoechst method

Materials

Buffer: 0.05 M $NaPO_4$, 2.0 M NaCl, 2 mM EDTA pH 7.4.

Hoechst reagent: 0.1 μg/ml Hoechst 33258 in buffer.

Standard DNA solution: 8 mg/ml of calf thymus DNA in distilled water.

Method

1. Homogenize or sonicate to lyse a cell suspension (10^5 cells/ml) in buffer.
2. Dilute lysate or standard DNA solution 1 in 10 in Hoechst reagent.
3. Measure fluorescence with an excitation λ of 356 nm and emission λ of 492 nm.
4. Determine DNA concentration by reference to standard DNA.

(b) DAPI method

Materials

Buffer: 5 mM HEPES, 10 mM NaCl pH 7.

DAPI reagent: A stock solution (× 100) contains 300 mg DAPI (4,6-diamidino-2-phenylindole) in buffer.

Standard DNA solution: 8 mg/ml of calf thymus DNA in distilled water.

Method

1. Homogenize or sonicate to lyse a cell suspension (10^5 cells/ml).
2. Dilute 150 μl lysed cell suspension with 850 μl buffer.
3. Prepare a DAPI solution (× 10) by diluting 100 μl of DAPI stock solution with 900 μl of buffer and mix well. Prepare a DAPI working solution by adding 0.5 ml of DAPI (× 10) to 4.5 ml of buffer.
4. Add 50 μl of DAPI working solution to each cell suspension or standard DNA (up to 0.8 μg) in a tube which is kept dark by a foil cover.
5. Vortex the tubes and let stand for 30 min.
6. Measure fluorescence with an excitation λ of 372 nm and emission λ of 454 nm.
7. Determine DNA concentration by reference to the standard DNA.

Notes

1. The DNA content of diploid cells is usually constant, although variations can occur as a result of the distribution of cells through the cell cycle. Cells in the Gl phase have the normal diploid content of DNA, which is typically 6 pg per cell.
2. DNA measurement is probably one of the best indicators of cell concentration in solid tissue.

Determination of the Glucose Concentration in Culture

(a) Glucose oxidase assay

Materials

Glucose oxidase/peroxidase reagent. Dissolve the contents of a reagent capsule from Sigma in 39.2 ml of distilled water. Each capsule contains 500 units of glucose oxidase and 100 units of peroxidase.

O-Dianisidine reagent. Dissolve the contents of a vial of o-dianisidine in 1 ml of distilled water. Each vial contains 5 mg of o-dianisidine dihydrochloride.

Assay reagent. Mix 0.8 ml of o-dianisidine reagent with 39.2 ml of glucose oxidase/peroxidase reagent.

Glucose standard solution. 1 mg/ml of D-glucose.

Sulfuric acid, 12 M.

Method

1. Start the reaction by adding 2 ml of assay reagent to glucose standard or culture media supernatant (0.01-0.1 ml). Make the assay volume up to 3 ml with distilled water.
2. Allow the reaction to proceed for 30 min at 37°C.
3. Stop the reaction by adding 2 ml of 12 M H_2SO_4
4. Measure the absorbance at 540 nm.
5. Determine the glucose concentration of the media samples against standard values obtained with the glucose solution.

Notes

Reaction 1 is catalyzed by glucose oxidase (GOD) and reaction 2 by peroxidase (POD). The dye, o-dianisidine hydrochloride is reduced by hydrogen peroxide to a product which has a pink colour in the presence of sulfuric acid (reaction 3) and is measured colorimetrically. The glucose oxidase kit from Sigma Chemical Co. contains glucose oxidase/peroxidase reagent and o-dianisidine reagent.

D-glucose + H_2O + O_2 $\rightarrow$ D-gluconic acid + H_2O_2 (1)

H_2O_2 + reduced o-dianisidine $\rightarrow$ $2H_2O$ + oxidized o-dianisidine (brown) (2)

oxidized o-dianisidine (brown) + H_2SO_4 $\rightarrow$ oxidized o-dianisidine (pink) (3)

(b) Hexohinase assay

Materials

Glucose (HK) assay reagent. Dissolve the contents of a reagent vial (Sigma) into 20 ml of distilled water. The dissolved reagent contains 1.5 mM NAD, 1.0 mM ATP, 1 unit/ml hexokinase and 1 unit/ml glucose 6-phosphate dehydrogenase.

Glucose standard solution. 1 mg/ml of D-glucose.

Method

1. Mix 10-200 μl of standard glucose solution or sample of culture media with 1 ml of assay reagent. Make the total assay volume up to 2 ml with distilled water.
2. Incubate at room temperature for 15 min.
3. Measure the absorbance at 340 nm.

4. Determine the glucose concentration of the media samples against standard values obtained with the glucose solution.

Notes

Hexokinase converts glucose into glucose 6-phosphate (G-6P) in the presence of ATP (reaction 1). The G-6P is immediately converted into 6-phosphogluconate by glucose 6-phosphate dehydrogenase (reaction 2). The associated formation of NADH is monitored by the change in absorbance at 340 nm and this is proportional to the concentration of glucose originally present. The hexokinase kit contains a hexokinase/glucose 6-phosphate dehydrogenase reagent. The kit includes a glucose standard solution (1 mg/ml).

D-glucose + ATP → glucose-6P + ADP (1)

glucose-6P + NAD → 6-phosphogluconate + NADH + H^+ (2)

The sensitivity of this assay can be increased by measuring the rate of increase of absorbance at 340 nm.This can be achieved with a recording spectrophotometer or using the kinetic mode of a multi-well plate reader.

(c) The glucose analyzer

Equipment

YSI model 27 Industrial Analyzer

Method

1. Fit the appropriate membrane into the analyzer for glucose analysis.
2. Calibrate the instrument with standard glucose solutions (2-5 g/l).
3. Inject 25 μl of a cell-free sample of culture supernatant into the instrument and compare with standard readings.

Notes

1. The sample is injected into a membrane which converts the glucose to hydrogen peroxide which can be determined by a sensor system based on a Clark electrode.The latter consists of a platinum electrode which measures the hydrogen peroxide amperometrically.

   $H_2O_2 \rightarrow 2H^+ + O_2 + 2e^-$ (1)

   $AgCl + e^- \rightarrow Ag + Cl^-$ (2)

   Current flow in the platinum anode is linearly proportional to the local concentration of hydrogen peroxide. This electrode is maintained at an electrical potential of 0.7 V with respect to a silver/silver chloride reference electrode, the potential of which is determined by the reaction above. The signal current which is

proportional to the quantity of injected glucose is converted to a voltage by the instrument circuitry.
2. This method of analysis is particularly suitable for the analysis of glucose in multiple samples of culture medium.
3. By the selection of the appropriate membrane in this instrument various analytes can be determined such as glucose, sucrose, starch, lactose, galactose, glycerin, lactate or ethanol.

Determination of the Viable Concentration of Cells by Dye Exclusion: Tetrazolium Assay

Materials

MTT reagent: 5 mg/ml of the tetrazolium dye, MTT (3-(4,5-dimethylthiazol-2-yl)-2,5-diphenyltetrazolium bromide) in phosphate buffered saline (PBS) pH 7.4.

SDS reagent: 10% w/v sodium dodecyl sulfate (SDS), 45% w/v N,N-dimethyl formamide in water adjusted to pH 4.5 with glacial acetic acid.

Method

1. Remove the media from adherent cells in a multi-well plate and add 0.1 ml MTT reagent Alternatively, add 0.1 ml MTT reagent to a 1 ml cell suspension in PBS.
2. Incubate for 2 h at 37°C.
3. Add 600 μl of SDS reagent and mix to solubilize the formazan crystals.
4. Measure the absorbahce at 570 nm.

Notes

1. The method is particularly convenient for the rapid assay of replicate cell cultures in multi-well plates. Plate readers are capable of measuring the absorbance of each well of a standard 96-well plate.
2. It is important to ensure that the coloured formazan salt formed from MTT is completely dissolved in the SDS reagent.
3. Alternatively, tetrazolium salts can be used in this assay such as XTT and WST I which are available from Boehringer-Mannheim. These form soluble coloured products.

Determination of Cell Viability by Lactate Dehydrogenase Determination

Method

1. Mix 2.8 ml Tris HCI (0.2 M) pH 7.3, 0.1 ml NADH (6.6 mM) and 0.1 ml sodium pyruvate (30 mM) in a cuvette.

2. Pre-incubate for 5 min at the desired reaction temperature (25°C or 37°C).
3. Start reaction by adding 50 μl of sample or standard LDH enzyme.
4. Record enzyme activity as an absorbance decrease at 340 nm.

Notes

1. This method is well suited for the determination of multiple samples, particularly if a multi-well plate reader is available.
2. Care must be taken when interpreting the results by this method because the LDH content per cell can change considerably during the course of batch culture.
3. The loss of cell viability can be expressed as the activity of LDH in the medium as a proportion of total LDH in the culture.

Determination of the Intracellular Energy Charge

Materials

ATP monitoring reagent/ATP-MR contains a lyophilized mixture of firefly luciferase, D-luciferin, bovine serum albumin, magnesium acetate and inorganic pyrophosphate. Re-constitute each vial with 4 ml buffer plus 1 ml potassium acetate (1 M).

ATP standards (LKB):ATP (0.1 μmol) and magnesium sulfate (2 μmol).

Buffer: 0.1 M Tris-acetate pH 7.75.

PK-PEP reagent 55 μl tricyclohexylammonium salt of phosphoenolpyruvate (0.2 M) + 50 μl pyruvate kinase (500 U/mg) in Tris buffer.

MK-CTP reagent 95 μl myokinase (2500 U/mg) + 10 μl CTP (110 mM) in Tris buffer.

Method

1. Extract soluble nucleotides by addition of 0.1 ml perchloric acid (20% v/v) to 1 ml of a cell culture sample (10° cells/ml).
2. Place on ice for 15 min and centrifuge for 5 min at 10,000 g.
3. Remove supernatant and neutralize with 5 M KOH.
4. For ATP determination: Mix 860 μl buffer, 100 μl ATP-MR and 10 μl sample.
5. For ADP determination: Add a further 10 μl PK-PEP
6. For AMP determination: Add a further 10 μl MK-CTP.
7. For standardization: Add a further 10 μl ATP standard.
8. Measure the light emission in a luminometer (for example, LKB 1250) after 1 min of each stage of addition.

Notes

1. The luminescence assay is dependent upon the emission of light resulting from the enzymic oxidation of luciferin, a reaction requiring ATP

   $ATP + LH_2 + O_2 \rightarrow AMP + PPi + CO_2 + L + light$

2. ADP and AMP can also be measured by the luciferase assay after conversion to ATP by coupled enzymic reactions.

   Pyruvate kinase:

   $ADP + PEP \rightarrow ATP + pyruvate$

   Myokinase:

   $AMP + CTP \rightarrow ADP + CDP$

3. The measurement of energy charge is more time consuming than the routine counting procedures discussed earlier but can allow a means of monitoring the decline in the energy metabolism of a cell culture that occurs during the loss of viability.

Determination of the Cellular Rate of Protein Synthesis

Method

1. Add ^{3}H-leucine or ^{35}S-methionine (Amersham) at a final specific activity of 20-40 μCi/ml to cell suspension at 5-10 $\times$ 10^6 cells/ml.
2. Remove 5-10 $\times$ 10^5 cells at each time point up to 4-6 h.
3. Isolate cell pellet by centrifugation in a microcentrifuge tube and wash in PBS.
4. Precipitate protein by addition of 500 μl trichloroacetic acid/TCAA (5%) containing unlabeled amino acids.
5. Wash the protein precipitate three times in the TCAA solution.
6. Add 30 μl NCS™ tissue solubilizer (Amersham) to the pellet and leave for 60 min.
7. Cut tip of tube and place in scintillation fluid for radioactive counting.

Notes

1. The cells should be incubated in the medium for sufficient time to measure radioactivity in the extracted cell pellet. Normally 4-6 h is sufficient but this may be longer.
2. The rate of DNA synthesis of a cell population can be determined in a similar assay to that described for protein synthesis but using a radioactively labeled nucleotide precursor such as trltiated thymidine (^{3}H-TdR) or deoxycytidine (^{3}H-CdR Amersham).
3. The exposure period may be short (30-60 min) for DNA synthesis rate determinations and a specific activity of 1 μCi/ml of culture is sufficient.

4. Higher specific activities may be required if using culture media containing the corresponding nonradioactive components such as methionine or thymidine.

Cell Characterization by Giemsa Banding of Chromosomes (Karyotype Analysis)

Materials

Colcemid (20 μg/ml).

Hypotonic solution (0.075 M KCI).

Fixative solution (methanol/acetic acid 3:1 v/v).

Staining solution (6.5 ml PBS, 0.5 ml trypsin 1%, 2.5 ml methanol and 0.22 ml Giemsa).

Method

1. Add colcemid (final concentration: 0.1-0.4 μg/ml) to a T-flask containing cells growing in exponential phase.
2. Remove the medium after approximately 6 h. For attached cells this can be by decanting.
3. Collect cells by centrifugation (for suspension cells) or by trypsinization and centrifugation (attached cells).
4. Allow the cells to swell in a hypotonic solution (0.075 M KCl) for 15 min at 37°C.
5. Collect cells by centrifugation, re-suspend in 0.075 M KCl and slowly add the fixative solution. Leave for 20 min.
6. Collect cells and re-suspend in the fixative.
7. Add the cell suspension dropwise onto a microscope slide and allow to dry.
8. Flood the slide with trypsin-Giemsa staining solution and leave for 15 min.
9. Observe the stained chromosome spreads microscopically. A photomicrograph will allow close inspection and analysis of the individual chromosomes.

Notes

1. Colcemid is a mitotic inhibitor and will block cells at mitosis. At this stage distinct chromosomes can be seen in the nuclei of cells.
2. The hypotonic solution causes cell swelling and maximizes separation between chromosomes.
3. The staining solution containing trypsin and Giemsa will break down associated protein and will stain the chromosomes with a characteristic pattern of G-bands.

7

Three-dimensional Culture Techniques

So far, the techniques described in this book have involved mainly the growth of dispersed cells either in suspension or in monolayer culture. Although these techniques have many advantages, the loss of structural integrity of the tissue may significantly alter the functional properties of these cells. Many general aspects of cells can be studied using this model, but the integrated function of the whole organism requires more complex models. As a result, culture methods have been developed in which the arrangement of cells resembles that *in vivo*.

These are three types of mediators of cell differentiation; soluble factors (such as hormones and paracrine factors); changes in the composition of extracellular matrix; and direct cell interactions (via gap junctions or surface information exchange). For the most part, these features are missing in dispersed cell culture and in this chapter we discuss more complex culture techniques in which much of the physiological environment is either retained or reconstructed.

In the first approach, the three-dimensional structure of the tissue is retained by explantation and growth of organs or parts of organs so that architecture is preserved and outgrowth of isolated cells is minimized.

The three-dimensional culture of tissue that is handled so that it retains some of the normal histological features of that tissue is called *organ culture*. An alternative approach is not to try to preserve the *in vivo* architecture of an organ, but to actually reconstruct the three-dimensional structure *in vitro*.

ORGAN CULTURE

Generally, organ cultures are more difficult to prepare than cell cultures. Once a cell line has been isolated and cloned, preparing fresh cultures simply requires inoculation of fresh medium. Organ cultures, however, cannot be propagated. Thus each experiment requires fresh tissue from the donor source. The production of such explants has to be done carefully in order to retain the structural integrity of this tissue.

Experiments using organ culture generally involve a large degree of experimental variation between replicates, so that reproducibility is less easily achieved than in cell culture. Organ culture is essentially a technique for studying the behaviour of integrated tissues rather than isolated cells and structural integrity is the main reason for adopting organ culture as an *in vitro* technique in preference to cell culture. As such, organ culture lies somewhere between cell culture on the one hand—and experiments using whole animals on the other—both in terms of the types of results which can be obtained and the complexity of the experimental model. As with all experimental models, caution should be exercised in the interpretation of results. Sometimes the systemic factors are important, for example a drug may be metabolized *in vivo* but not *in vitro*, thus giving qualitatively and quantitatively different results in the two situations. There is also time limitation on organ culture. Since these cultures cannot be maintained indefinitely any effects which appear only after several months will not be observed using this model.

When cells are cultured as a solid mass of tissue, a number of problems arise which are not an issue with cell culture. In particular, gaseous diffusion and the exchange of nutrients become limiting. To overcome these problems, a number of different techniques have been developed. Each of these have their limitations.

Historical Development of Organ Culture

As stated already that one of the earliest tissue culture experiments was performed in 1885, when Roux maintained the medullary plate of a chick embryo in warm saline for a few days. In 1897, Loeb was the first to culture fragments of adult rabbit liver, kidney, thyroid and ovary on small plasma clots inside a test tube. He found they retained their normal histological structure for three days.

It is now known that the best results are obtained by growing the tissue on a solid support within a fluid medium. The organ culture is kept at the interface between liquid and gaseous phases to facilitate

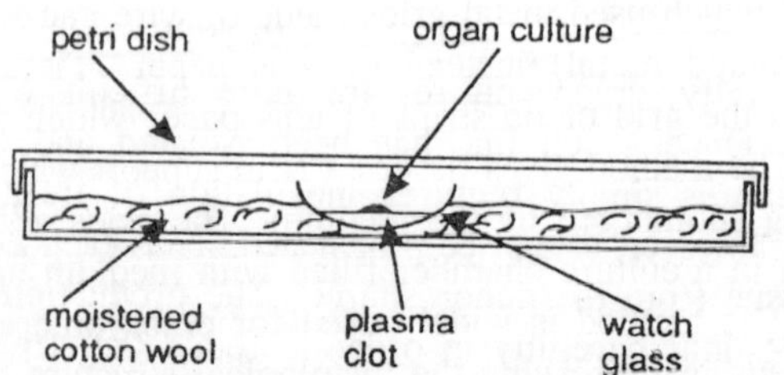

Fig. 7.1. Diagram to show the watchglass technique of organ culture.

gas exchange while retaining access to nutrients. There has been a steady development of new and improved methods to provide these conditions. The first substrate used was a clotted plasma substrate. In 1929, Fell and Robinson introduced the '*watchglass technique*', by which organ rudiments or organs were grown on the surface of a clot consisting of chick plasma and chick embryo extract contained in a watchglass.

This became a standard technique for morphogenetic studies of embryonic organ rudiments and the method was later modified to investigate the action of hormones, vitamins and carcinogens in adult mammalian tissues. The main disadvantage of this technique is that explants often liquify the clot and sink into a pool of liquified medium. For this reason the technique was superseded by the agar clot technique, which uses agar instead of clotted plasma, in the 1940s.

The next major development occurred in 1954, when lens paper (used for cleaning microscope lenses) was used as a floating raft on which cultures were grown. It was possible to study cell-cell interactions by growing different types of tissue on either side of the lens paper. The use of rafts floating on a fluid medium had a number of disadvantages. The rafts often sank and the tissues were frequently immersed in the medium. This difficulty was overcome by the '*Grid*

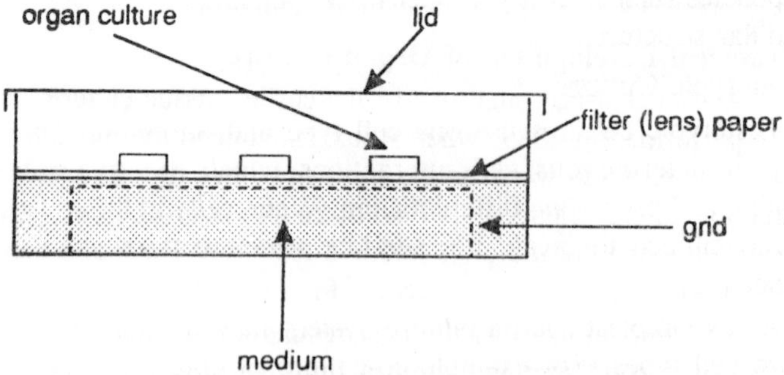

Fig. 7.2. Diagram to show the grid technique of organ culture.

technique' which used metal grids made of wire gauze (and later more rigid expanded metal) instead of lens paper. Tissues are cultured directly on the grid or on strips of lens paper which are deposited on the grid. The method gives tissues a firm support which prevents them from being submerged by the medium. The grids with their explants are placed in a culture chamber filled with medium up to the level of the grid. The method is widely used for embryonic and adult tissues and succeeds in preserving the histological structure of the tissues.

Reconstruction of Three-dimensional Structure

In organ culture, organs or parts of organs are grown in culture so that they retain the histological features of the tissue *in vivo*. There are other culture methods which start with dispersed populations of cells but encourage the arrangement of these cells into organ-like structures. In other words rather than preserving the architecture of an organ, the three-dimensional structure is actually reconstructed *in vitro*. In these systems important physiological factors may be quantified thus overcoming some of the disadvantages of organ culture.

These types of culture can be sub-divided into two basic types:

1. Histotypic cultures;
2. Organotypic cultures.

Histotypic Culture

Culture systems using one characterized cell line propagated at high density in the presence of appropriate extracellular matrix and soluble factors are called *histotypic*. For example, vascular endothelial cells will form capillary tubules in the presence of appropriate soluble factors when grown in a collagen matrix. Homologous cell interactions occur because the cells are cultured at high density. An alternative approach is the use of cellulose sponges coated with extracellular matrix components such as collagen. Cells may penetrate the sponge and form glandular structures.

Organotypic Culture

Histotypic cultures use one cell type and, therefore, exclude the study of heterologous cell interactions which are so important in initiating and promoting differentiation. When cells of different lineages are recombined to create a tissue-like structure the culture is called organotypic.

At its simplest such a culture system may involve the *co-culture* of two cell types. For example co-culture of epithelial and fibroblast cell clones derived from the mammary gland allows the epithelial cells to differentiate functionally as demonstrated by their ability to produce milk proteins in the correct hormonal environment. This functional differentiation is preceded by the formation of characteristic structures such as three-dimensional cords in which fibroblast cells organize themselves into bundles which are then enveloped by epithelial cells.

As well as applications in research, the generation of organ-like structures from dispersed cell lines may one day be used in the treatment of human disease. In 1989 the first working '*organoid*' was reported—an artificial liver surgically implanted into the peritoneal cavity of a rat. The organoid was made from a combination of cells, growth factors, collagen and Gore-tex fibres. Such organoids could potentially replace diseased organs or deliver genetically altered cells into a patient's body.

Categorizing Tissue Culture

You may have noticed that the different categories of tissue culture methods defined in this book are actually overlapping. This reflects the range of experimental approaches to the *in vitro* cultivation of animal cells. Although it is convenient to use such a classification it is important to recognize that there are overlaps between these methodologies.

Uses of Embryonic Tissue

It has already been described that the cell cultures can be derived from embryonic tissue. In general, cell lines from embryonic tissue survive and grow better than those from adult tissues. Cultures of rodent embryo cells routinely give rise to continuous cell lines whereas Hayflick's experiments using human foetal lung fibroblasts showed that human foetal cells have only a limited capacity to grow in culture, since the cultures invariably degenerate. Nevertheless, human embryonic cells in culture have had a number of uses including the generation of vaccines for mass vaccinations against polio.

A more recent development has been the removal of embryonic stem cells (ES-cells) from the embryo during the blastocyst stage of development. These cells can be grown in culture for many generations and are of particular interest because they can be manipulated in culture and then re-introduced into embryos.

In addition, much organ culture has been performed using embryonic tissue. Organ rudiments can develop in vitro in a similar manner to *in vivo*. The earliest tissue culture experiments were performed using the nervous tissue of chick embryos (Roux's experiments) and many of the studies of early development were carried out using organ culture. The use of embryos is much broader than being merely sources of cells for cell and organ culture. In fact embryos are of tremendous importance.

EMBRYO CULTURE

These days, whole embryos can be grown *in vitro* and a high proportion of these embryos will develop well in culture. The term embryo culture is used to describe the *in vitro* development or maintenance of isolated embryos. This term excludes cell culture and

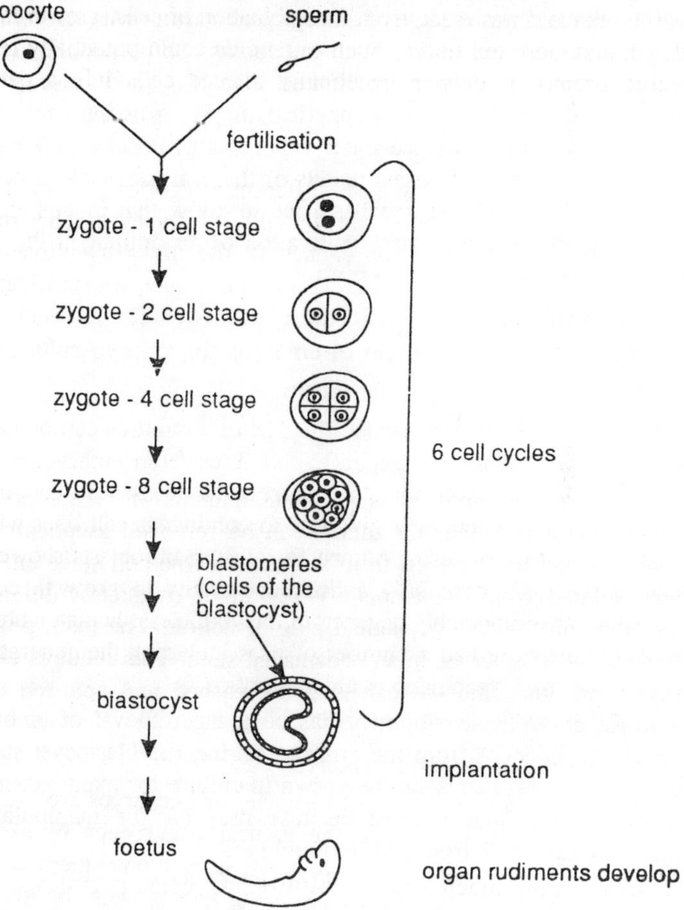

Fig. 7.3. Stages in the development of a mammalian embryo.

organ culture, and refers simply to the cultivation of intact embryos *in vitro*. The word *embryo* has been used to mean different things by different people. Clearly the word refers to a stage of development before birth. The broadest definition of the word embryo may be applied to structures that exist starting with the fertilized ovum and ending with birth. Strictly speaking, however, this is incorrect.

An embryo is a spatially defined entity that can develop directly into a foetus. In the foetus, the main structures of the body are formed, so the word embryo implies that all the main structures of the body have not yet been formed. Thus the word embryo is often used to denote the developmental stage of an organism starting with the fertilized ovum (the *zygote*) and ending with the foetus. Some people say that the term embryo is even more limited than this, and refers to the stage between the blastocyst (a spherical mass consisting of a single layer of cells from which the inner cell mass projects into a cavity) and the foetal stages. The exact details of these arguments are beyond the scope of this text, and it will suffice to know that in this chapter we are referring to the *in vitro* cultivation of an animal in the early stages of development, prior to birth.

Sources of Embryos

There are two main sources of embryos for embryo culture:

(i) by dissection;

(ii) by *in vitro* fertilization (IVF).

Dissection of embryos

One source of embryos is pregnant animals—for example if a pregnant mouse is killed, the uterus can be removed aseptically and the embryos can be retrieved from the uterus. One cell stage embryos can be isolated from the ampullary region of oviducts from mated female mice on the date of vaginal plug detection. The main problem is the removal of embryos in an undamaged state. This requires careful dissection and the embryos can then be washed and cultured under silicone oil at 37^{0}C.

In vitro fertilization to obtain embryos

In invertebrates vast numbers of gametes and embryos are available for experimental studies due to the natural occurrence of external fertilization. The other source of embryos is to create them *in vitro* by fertilizing oocytes from the female using sperm from the male, in other words by *in vitro* fertilization (IVF). The technique of IVF was developed primarily in mice and hamsters to study the process of fertilization and early development. In mammals, minute numbers of eggs and embryos are produced in each cycle by females. However, females can generally be induced to superovulate using gonadotrophins.

Fertilized ova can be grown to the blastocyst stage *in vitro*, and the embryos apparently undergo normal development since re-implanted embryos can go on to form healthy animals. The technique of IVF has been applied to a large number of species during the past 25 years with varying degrees of success. The ability to induce superovulation depends upon knowledge of the hormonal control of the oestrous cycle. The most common way is to use treatments with progesterone and injections of gonadotrophin. The exact treatment is dependent upon the species of animals being used.

Production of Embryos *in vitro*

Induction of superovulation

In many mammals only a single follicle or a few follicles in the ovary develop to produce mature oocytes. Yet in most mammalian ovaries there are many follicles each capable of producing oocytes. Three types of follicles can be distinguished. Some are large (10 mm in diameter), some are medium in size (2-10 mm), while the remainder are small (less than 2 mm). Often 50-80 follicles of diameter 2-10 mm develop during the normal oestrous cycle. However, all but a few degenerate. If the animal is injected with gonadotrophic hormones such as *follicle stimulating hormone* (FSH), the development of several follicles may be maintained. In this way, many oocytes may be released during ovulation. These superovulation procedures, however, still suffer from highly variable yields.

One of the alternatives to superovulation is the *in vitro* maturation and fertilization of oocytes. The method that yields best results is to use transvaginal ultrasound guided puncture of follicles. This uses ultrasound to guide a needle via the vagina into the follicles to extract the oocytes. The method is, however, generally only applicable to larger animals such as cattle. The oocytes are then matured *in vitro*.

In vitro maturation has the attraction that immature oocytes can be collected repeatedly from the same animal irrespective of the day of the oestrous cycle. It is becoming common practice to use a combination of these two strategies. Superovulation is induced by gonadotrophic hormone treatment and the oocytes are normally collected 2-4 hours before ovulation is expected. They are then allowed to mature *in vitro* for 1-6 hours before insemination. It is much easier to locate

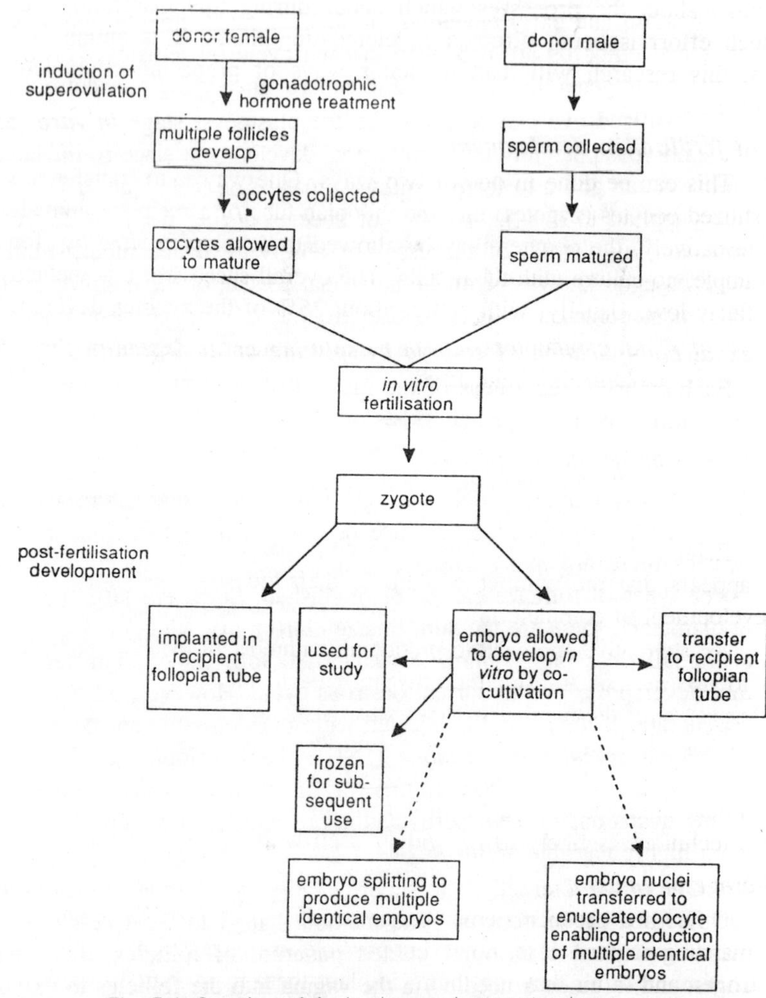

Fig. 7.4. Overview of the in vitro production of embryos.

the follicles before ovulation (and therefore the oocytes) than it is to locate oocytes released after ovulation.

Fertilization in vitro

Fertilization is achieved *in vitro* by the addition of sperm. Often the sperm has to be pre-treated to make it competent in fertilization. For example, ejaculated bovine sperm is not able to fertilize the oocytes *in vitro*. It first needs to be treated with heparin or other substances like caffeine and lysophosphatidyl-choline to become mature. Little is known about the processes which occur during this maturation but much effort is being directed to elucidating them. It is anticipated that this research will lead to better ways of preparing competent sperm.

Post fertilization development

This can be done in one of two ways. One way is to transfer the fertilized oocytes (zygotes) into the fallopian tube of a recipient animal. Alternatively, the oocytes may be allowed to develop *in vitro* by, for example, co-culture with tubal cells. The overall success of this method is fairly low. Usually, with cattle, about 25% of the zygotes develop.

Asexual multiplication of embryo by splitting or nuclear transfer

Each embryo, regardless of the method of production, contains a unique combination of the genotypes of its parents. Identical embryos can be produced by embryo splitting. In this technique, early stage embryos are first halved and then quartered. Each quarter is still capable of developing into a viable young.

Efforts to divide an embryo into eight parts have not succeeded. It appears that the number of cells is then too small to support the development of complete new embryos.

An alternative approach to producing a number of identical embryos is to use nuclear transplantation. Since each nucleus of an embryo, irrespective of the number of cells, has the same chromosomes present, each has the potential to support complete embryo development. If these nuclei are transplanted into enucleated oocytes, then multiple, identical embryos should be produced. Although still in its infancy, this technique is likely to be widely used with domestic animals.

Storage of embryos

Obviously, producing embryos in vitro is expensive and a technically demanding process. It is, however, possible to freeze embryos in liquid nitrogen and, after thawing them, implant them many months or years after fertilization. Since 1983, it has become possible to freeze and store human embryos in this way. However, freezing excess embryos is prohibited in some states and countries.

The zygotes produced by *in vitro* fertilization may be used in a number of different ways. If they are allowed to develop into embryos *in vitro*, then they may be used to study the early stages of embryo development. The embryos may also be transferred to a suitable host to give rise to foetuses and hence to mature offspring. Alternatively, these embryos may be split to produce multiple embryos for use in research or to produce multiple identical offspring. In the latter case the 'split' embryos are transferred to suitable hosts. This technique has been successfully used to produce identical farm animals but ethical and legal considerations exclude its application to humans.

Evolution of embryo quality

A variety of criteria may be used to evaluate the quality of embryos that are to be used for transfer or cryopreservation. This criteria may include items from the following list:

- has the embryo achieved the right levels of development it should have reached at the time of harvesting?
- are the blastomeres (cell of the blastocyst) uniform?
- are the membranes distinct and intact?
- are vesicles present?
- are there signs of degeneration and dissolution?

On this basis embryos may be classified as excellent, good, fair or poor. In other words, these criteria are qualitative and rather imprecise.

Ratings, based on physiological parameters, have also been used. These include dye-exclusion tests and measurement of key enzymes. These tests are, however, invasive.

Much effort is being put into developing non-invasive techniques. Recently, a non-invasive technique has been developed based on the changing metabolic requirements of embryos. In mice, it has been shown that pyruvate is an essential nutrient during the first cleavage division. Glucose is unable to support development until the 8-cell stage. Thus, using microfluorescence-enzymatic methods to determine small changes in the concentrations of pyruvate and glucose in microdrops of media used to cultivate embryos provides a basis for evaluating their quality. It appears that pyruvate requirements of human embryos are similar to those of mouse embryos, so this test may also be used to evaluate the quality of human embryos.

Applications of IVF and Embryo Transfer

The practical applications of IVF and embryo transfer can be studied under four headings:

Applications in Embryology

The development of the mammalian embryo in the protected environment afforded by the female reproductive tract presents a natural obstacle to research. Thus the use of in vitro techniques for the study of early mammalian development is easier and more informative than attempting to study the events in the natural environment. Manipulation of embryos is an important tool in the experimental embryology of mammals and embryos have been used to provide information about normal development, growth and differentiation and the influence of extrinsic factors such as toxic agents in teratological studies. Much has been learnt about the mechanisms which control normal mammalian development using the animal embryo as an experimental model and these animal models have provided information which had been relevant to understanding human development.

Applications in Transgenics

In 1980, results of experiments were published in which DNA was micro-injected into the male pronucleus of fertilized mouse eggs. The injected eggs were re-implanted into pseudopregnant females to develop, and it was shown that the micro-injected DNA had become incorporated into the chromosomes of the resulting newborns. Although no expression of this incorporated DNA was detected, subsequent studies reported high levels of expression of micro-injected fusion genes using similar techniques. Integrated genes were shown to be transmitted through the germ line and progeny were also shown to express the genes. Animals which contain integrated foreign DNA are generally called transgenic and the foreign genes are conventionally referred to as transgenes. There are countless applications of transgenic methodology including research into genetic diseases and improvements in productivity of animal livestock.

Application in Animal Breeding

The technique of IVF followed by embryo transfer is transforming animal breeding since it facilitates sex selection (for example allowing the selection of only females for dairy herds) and intercontinental shipping of embryos in place of adult animals. The transfer of bovine embryos from prize cows, each worth thousands of pounds, has become a significant industry.

The sexing of embryos is usually based on the detection of sex-specific antigens on the surface of the cells. Alternatively sex-specific DNA probes may be used to detect sex-specific nucleotide sequences. Although this technique is reliable, a single blastomere still needs to be removed from embryos for testing purposes.

The approach of using embryos for experimental objectives is full of controversy and it is important to be aware of the relevant legislation before performing any experiments. For example, recent changes in the UK mean that use of embryos and foetuses beyond 50% gestation or incubation comes under the Animal Experiments Act of 1986. You should be acquainted with the legal status of embryo culture in your country if you plan to use this technique.

Applications in Human Reproductive Technology

In recent years there has been a renewal of interest in IVF procedures stimulated in part by the successful use of this technique for the treatment of human infertility. The first child conceived with the help of the IVF technique was born in July 1978, and since then the IVF technique has led to the birth of almost 1000 babies (popularly called '*test tube babies*') worldwide.

In human ovaries at birth there are about 2 million cells which can mature into potential egg cells (ova). The complete maturation of these oocytes requires hormonal signals and generally one such maturation will occur in a monthly cycle. All other potential oocytes remain in a resting state or degenerate. The original steps in the maturation process occur in follicles within the ova. Follicles become fluid-filled, expand to about 2 cm in size and bulge out of the surface of the ovary. When the follicle ruptures, the maturing oocyte is propelled towards the fallopian tube. At this stage the cell is called a secondary oocyte. The maturation of this secondary oocyte into an ovum with the correct amount of genetic material is completed when the spermatozoan penetrates the membrane of the cell. When the maternal and paternal chromosomal materials mix, the zygote is produced. The first cell division produces 2 blastomeres, which may separate to produce identical twins. Within a few days after fertilization, the ball of dividing cells has travelled through the fallopian tube and enters the uterine cavity.

In some individuals, however, the events between ovulation and implantation of the embryo in the uterine cavity are restricted or prevented. For example, if the fallopian tubes are blocked, the oocyte cannot enter the tube and thus fertilization cannot be achieved.

These, and similar problems, may be overcome using IVF and in vitro embryo culture.

An early step in IVF is the removal of the oocytes from their follicles by laproscopy. The oocytes are normally aspirated 2-4 hours before ovulation is expected and allowed to mature *in vitro* for 1-6 hours before insemination. More than 85% of pre-ovulatory oocytes from patients with tubal disorders are successfully fertilized *in vitro*. The embryos are allowed to develop *in vitro* to the blastocyst stage and then re-introduced into the female for implantation, where about 20-30% of the replaced embryos result in pregnancy.

In order to increase the efficiency of treatment, the IVF programme generally includes ovarian stimulation to initiate maturation of more than one oocyte (superovulation), The woman is treated with a combination of clomiphene and human menopausal gonadotrophin (HMG) which act as follicle stimulants. Doctors use these oocytes to grow several human embryos *in vitro* and usually only some of these are used for implantations. Upto 10 embryos have been replaced on a single occasion. The embryos not implanted can be discarded, frozen down for future pregnancies, or used for research.

A number of techniques have been used to assess human embryo viability. These are either invasive (resulting in destruction of the embryo or parts of the embryo) or non-invasive (for example observations on the rate of cell division). Since chromosomal defects have been observed in otherwise apparently normal 8 cell embryos, many would argue that more invasive tests should be performed to avoid implanting embryos with genetic defects.

Experimental research on early human embryos is an emotive issue and the development of the IVF technique has raised many ethical problems. A public debate has centred on the question of whether research using human embryos should be permitted at all, and many people consider that it is unethical to create a human embryo for any purpose other than enabling a child to be born. Others argue that the possible benefits far outweigh any such considerations and contend that human embryos provide unique models for research into:

1. Diagnosis of genetic disease;
2. The diagnosis of infertility;
3. Improved treatments of infertility by IVF;
4. The causes and treatment of congenital malformations.

The ethical and legal issues surrounding human embryo research are subjects of major controversy. We do not intend to take a moral stand here. We must, however, alert you to the fact that work in this area is increasingly becoming subject to legislation and guidelines. Perhaps in this area of scientific activity more than any other emotions and fears run high and there is great disparity between the positions adopted by different countries and states. These reflect religious and cultural differences. What is acceptable in one state/country may be unacceptable elsewhere.

In the UK, for example, the legal and ethical issues raised by IVF and the possible research which could be performed on unused human embryos were made the subject of a committee of inquiry set up by the Department of Health and Social Security and chaired by Dame Mary Warnock. The Warnock Committee report (1984) makes a number of specific recommendations, one of which is that research on human embryos should be permitted upto the 14th day after fertilization. However, the issue of whether such studies are acceptable has not yet been resolved and, in the UK, several attempts have been made to completely prohibit embryo research.

The delay in the resolution of this problem reflects the difficulty of generating consensus views and the rapid development of these technologies. Some would argue that research should be allowed to continue so that the full capabilities of these techniques are assessed before legislation is formulated. In this way, we would be aware of the potential benefits (and potential dangers) before the imposition of regulations. Others would argue that this research is totally unethical and should be banned immediately. Of course, this would remove the moral and practical problems that might arise from these activities. At the same time, it would prevent the realization of the benefits that might be accrued.

Individuals working in this area, will, of course, come to their own conclusions. Nevertheless, it is important for researchers to check that all proposed experiments fall within the appropriate laws of their country. For any experiments using human material, there may be legislation concerning consent from hospital ethical committees, attending physicians or surgeons, and the 'parents' of the embryo.

Future direction of human embryo research

The development of IVF procedures has facilitated further treatments of infertility, including the possibility of donating eggs or embryos to women who are unable to produce their own oocytes because of primary ovarian failure or women who are known to carry genetic defects. This allows the male to participate genetically and the female to carry a normal pregnancy. The implantation of a fertilized egg into the uterus of another woman, is called surrogate parenthood. In 1987, a 48-year old grandmother successfully carried 3 foetuses from her own daughter's oocytes.

In addition, the removal of the early steps of reproduction from the setting of the woman's body has opened up new possibilities for parents which include:

1. Selecting the timing of birth. Foetal chromosomal disorders increase 2-3 fold when women are over 40 compared with those just a few years younger. Now that methods of freezing and thawing embryos are available, the time between oocyte retrieval and implantation may be extended allowing parents to choose the time of pregnancy;
2. Determining the sex and number of children;
3. Carrying our pre-implantation diagnosis and gene replacement;
4. Selection for specific traits.

Again, these new options are accompanied by a myriad of ethical, legal and social concerns.

8

Maintenance of Culture

This chapter considers some of the procedures for establishing cells in culture and the environment necessary to maintain favourable growth conditions.

How to Culture Cells and What to Expect

Inoculation

Once cells are obtained from a culture collection or isolated from animal tissue, a culture can be initiated by inoculating the cells into sterile growth medium. In order to obtain a reasonable growth rate, cells should be inoculated at a density of 10^4-10^5 cells/ml. This will allow growth in a simple batch culture to around 10^6 cells/ml or to 10^5 cells/cm^2 on a solid surface in about 3-4 days. The increase in cell concentration in a culture can be expressed as the multiplication ratio, which is defined as C_f/C_i, where C_f is the final concentration and C_i the initial concentration. Expect a 10-fold increase in cell concentration in a simple batch culture.

Anchorage-dependent cells attach to an available growth surface within a few hours of inoculation. The attachment process involves the flattening and spreading of cells into a characteristic shape.

Cell growth stops because of a nutrient limitation, an accumulation of a toxic metabolite or a lack of growth surface (for anchorage-dependent cells). If the medium is totally or partially replaced after 2 days, the maximum cell density may be higher.

Subculture

When cells stop growing in culture, new cultures can be established by inoculating some of the cells into fresh medium. This is called subculturing or passaging. It is important to subculture cells within a day or so of the maximum cell density to ensure continued growth in a new culture. Cells will lose their viability if they are left for too long before subculture. For cells grown in suspension, subculture involves dilution of the high-density culture with fresh medium. Dilutions from one in two to one in ten v/v would be suitable. It is advisable to pre-incubate growth medium in an incubator before inoculation. This ensures that the new culture is at the optimal temperature (37°C) and pH (~7.4) for maximum cell growth.

The subculture of anchorage-dependent cells involves detachment of the cells from the growth surface (substratum) of one culture flask and re-inoculation of the cells into fresh medium contained in a new culture flask. The cells are detached from their anchor by the process of trypsinization. Trypsinization was first introduced in the early days of cell culture but was not fully developed until the 1950s when its use enabled the establishment of homogeneous cell populations.

A proteolytic enzyme such as trypsin will break down the proteins that bind the cells to the culture surface. The trypsin is added to washed cells in the culture flask for a short period that is long enough to dislodge the cells from the substratum but not too long to damage the cells. Cells will detach from a surface by treatment with trypsin for 15 minutes. It helps to bang the edge of the T-flask with your hand. Detachment of cells by trypsin is more efficient if magnesium and calcium ions are removed from the medium. This is achieved by combining trypsin with EDTA ('versene') administered in a Ca- and Mg-free salt solution. EDTA is a chelating agent that may be used alone for removal of weakly adherent cells. The action of trypsin is stopped by the addition of serum-supplemented medium and centrifugation to remove the cells. The excess protein present in the medium serves to reduce the trypsin activity. If the cells are grown in serum-free medium a soybean trypsin inhibitor can be used.

Cultures are given a passage number which indicates the number of subcultures performed since the cells were obtained or isolated. The relationship between the passage number and the generation number ('population doublings') depends upon the 'split ratio', which is the number of new cultures established at each stage of subculture. A typical cell population will double in one day.

The simplest case occurs where a confluent culture is sub-cultured into two new cultures i.e. at a split ratio of 2. In this case, the generation number and the passage number are the same.

Phases of a Culture

The culture will follow a growth pattern. Several phases of culture can be identified from this pattern.

The lag phase

This is an early phase in which there is no apparent increase in cell concentration. This phase is associated with the cellular synthesis of growth factors which may be required to reach a critical concentration before growth takes place. The length of this phase is dependent upon the culture medium formulation as well as the initial concentration and state of the cells. The lag phase tends to be longer at low inoculation densities or if the viability of the inoculated cells is low. This may occur if subculture is delayed. If a high density of cells with good viability is inoculated into the culture medium then the lag phase may be eliminated altogether. Transformed cells have a lower requirement for growth factors and often show no lag phase ever when inoculated at lower concentrations.

In some circumstances, there may be a requirement to inoculate at a low cell density, for example during cell cloning. Cloning is the establishment of a culture from a single cell. This ensures that all cells in the culture are identical, i.e. a homogeneous population. In this situation cell growth can be enhanced by adding a feeder layer which consists of irradiated cells incapable of growth but metabolically active and capable of releasing growth factors into the medium.

The growth phase

During the exponential growth phase cells go through the cell cycle:

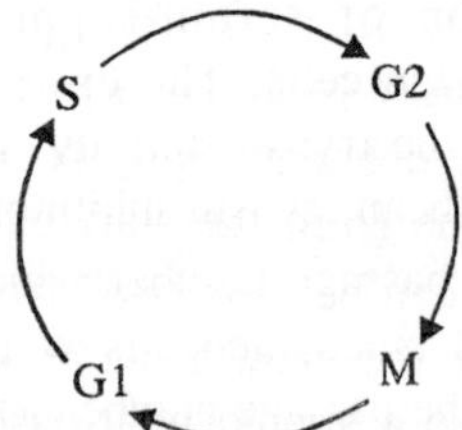

This phase involves an exponential increase in cell number which can be represented by the following equation:

$$N = N_0 . 2^X$$

or

$$\log_{10} N = \log_{10} N_0 + X . \log_{10} 2$$

where N = final cell concentration, N_0 = initial cell concentration and X = number of generations of cell growth.

The number of generations (X) for a cell inoculum of 10^5 cells/ml which increases to a density of 10^6 cells/ml can be calculated as follows:

$$\log_{10} 10^6 = \log_{10} 10^5 + X.\log_{10} 2$$

Therefore:

$$X = \frac{(\log_{10} 10^6 - \log_{10} 10^5)}{\log_{10} 2} = \frac{1}{\log_{10} 2} = 3.32 \text{ generations}$$

The doubling time during cell growth can be calculated from the equation:

$$t_D = \frac{T}{X}$$

where t_D = doubling time, T = total elapsed time and X = number of generations.

Thus, for a culture in which an initial cell density of 10^5 cells/ml reaches 10^6 cells/ml in 3 days the average doubling time is 0.904 days or 22 h.

Animal cells normally exhibit a doubling time of between 15-25 h during the exponential growth phase.

The specific growth rate (μ) is another growth parameter which is often calculated. This is a measure of the rate of increase of cell number (or biomass) at a certain cell concentration.

$$\mu = \frac{dN}{dt} \cdot \frac{1}{N}$$

or

$$\ln N = \ln N_0 + \mu \cdot t$$

where N_0 = initial cell concentration, N = cell concentration at time and t = elapsed time from the start of cell growth.

The value of μ (h^{-1}) can be calculated from the slope of a plot of ln N against time, t. For the growth curve, the specific growth rate is 0.032 h^{-1}. There is also a simple relationship between the specific growth rate and doubling time,

If $N = 2 \times N_0$, then $\ln 2 = \mu.t = \mu \times$ doubling time (t_D). Therefore:

$$\mu = \frac{\ln 2}{t_D} = \frac{0.6931}{t_D}$$

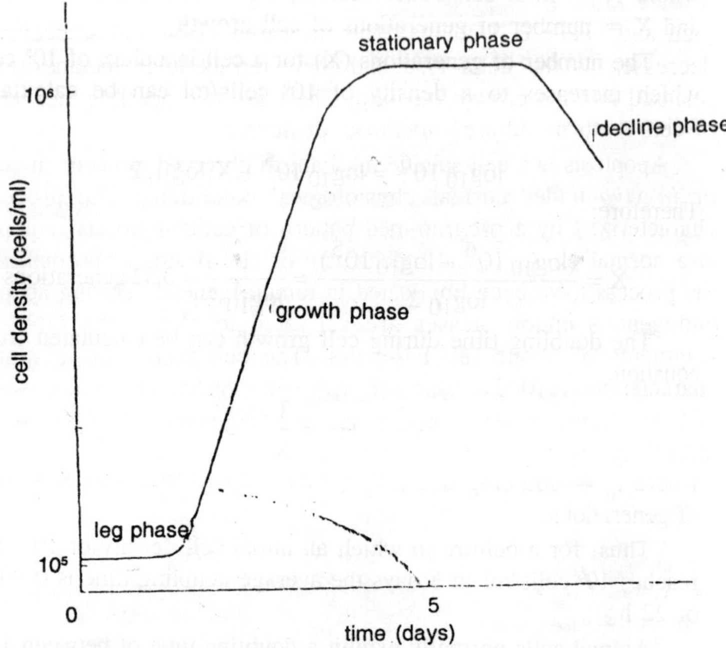

Fig. 8.1. Cell growth in culture.

The stationary phase

This phase occurs when there is no further increase in cell concentration. During the stationary phase, death rate = growth rate. At this point cell growth is limited by one of a number of conditions:

1. Nutrients may have been depleted to a level that cannot support further cell growth;
2. The accumulation of metabolic by-products may have reached a level which is inhibitory to cell growth;
3. The cells may have formed a complete cover over the growth surface. Growth may stop when a single monolayer of cells covers the available substratum (called '*confluence*'). This phenomenon is associated with cell-cell interaction.

During the stationary phase the cells may be metabolically active even though growth is not occurring. For example, in some cases high productivity of secreted proteins may occur during this phase.

The decline phase

This phase occurs as a result of cell death. Cell viability is lowest during the decline phase of culture—shown by a large difference between total and viable cell counts. The measured viable call concentration decreases as the cells lyse and their intracellular metabolites are released into the growth medium. There are two possible mechanisms of cell death in culture—*apoptosis* or *necrosis*.

Apoptosis is a cell suicide mechanism observed to occur in culture or *in vivo* under normal physiological conditions. The process is characterized by a programmed pattern of cellular events. Apoptosis is a normal physiological mechanism of cell death. Abnormalities in the process have been implicated in tumourigenesis. During apoptosis endogenous endonucleases are activated to cleave the DNA into fragments of about 180 basepairs. The cell membrane assumes a characteristic ruffled appearance with many blebs. This is followed by cell shrinkage, nuclear condensation and fragmentation of the cell into discrete membrane-enclosed apoptotic bodies.

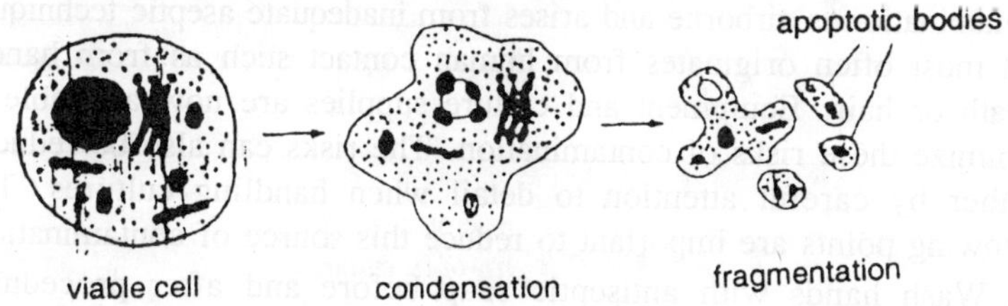

Fig. 8.2. The stages of apoptosis.

In vivo, the apoptotic bodies generated by cell breakdown are phagocytosed by adjacent cells. Apoptosis is thought to he important *in vivo* in regulating the proliferation of certain cell types for example, following immunological stimulation of lymphocytes. In cell cultures apoptosis may be initiated by the depletion of nutrients. This causes the cells to go into a decline phase rapidly with a limited stationary phase. This is observed particularly with certain cell types for example, lymphocyte hybridomas.

The alternative mechanism of cell death is necrosis, which follows cellular injury. This does not involve the cell fragmentation characteristic of apoptosis and the loss of cell viability in culture is relatively slow. Necrosis is a passive process that normally occurs when cells are subject to sudden severe stress. This is characterized by a breakdown of the plasma membrane leading to cell swelling and eventual cell rupture.

Importance of Aseptic Techniques

The major cause of failure and frustration associated with the operation of cell cultures is the problem of contamination. Cultures are susceptible to overgrowth by bacteria because of the difference in growth rates. Typical doubling times are:

1. Animal cell = 24 h
2. Bacteria = 30 min.

The slow growth rate of animal cells makes the culture particularly vulnerable to the fast growth rate of potential microbial contaminants such as bacteria and fungi. The safeguard against this problem is the meticulous use of standard aseptic-microbiological techniques for all transfer operations.

How to Prevent Contamination

The aseptic techniques necessary for cell culture can now be performed easily in a well-designed laboratory and are far less elaborate than those originally conceived by Carrel. The major source of contamination is airborne and arises from inadequate aseptic techniques and most often originates from human contact such as from hands, breath or hair. Equipment and culture supplies are now available to minimize these risks of contamination. The risks can also be reduced further by careful attention to detail when handling cultures. The following points are important to reduce this source of contamination.

1. Wash hands with antiseptic soap before and after procedures involving handling cells. Latex gloves may be worn.
2. Limit access to the laboratory when experiments are in progress.
3. Decontaminate working surfaces before and after each procedure.
4. Use a sterile space (e.g. laminar flow cabinet) for all manipulations.
5. Ensure that there is a barrier between the operator's face and the culture. This could be a face mask or more usually the perspex front of a laminar flow cabinet.
6. Use presterilized plastic culture flasks or ensure the adequacy of an inhouse sterilization operation. The plastic culture flasks are only used once.
7. Purchase media and serum from a reputable supplier to ensure contamination does not arise from this source. In the past, animal serum has been a source of mycoplasma infection but all batches of serum are now tested by the major suppliers and should be certified as mycoplasmafree. Mycoplasmas are Gram-negative bacteria (0.3-0.5 μm) capable of growth within the animal cell cytoplasm. This type of contamination is very difficult to detect in cell culture.

Media prepared in-house can be checked for sterility by incubation of a sample in a variety of broths or nutrient plates. Two commonly used isolation media suitable for revealing bacterial or fungal contamination are fluid thioglycollate medium and soybean-casein digest, both of which are available as dehydrated media. Samples can be inoculated into these media and contamination is revealed by increased turbidity. Incubation times for each sterility test should be at least 7 days.

The incorporation of antibiotics such as penicillin and streptomycin into cell growth media became common in the 1950s. Penicillin and streptomycin prevent the growth of Gram-positive and Gram-negative bacteria. Biological fluids and embryo extracts are particularly vulnerable to bacterial contamination. Supplementation with antibiotics reduced this problem. This is an option favoured by many laboratories. However, there is a dagger that the use of antibiotics may mask a contamination for a short time without eliminating it. This has the effcct of delaying a problem rather than removing it. For this reason antibiotics are not used as extensively as in the past. If your aseptic technique is good, then antibiotics should not be required.

How to Recognize a Contaminated Culture

Bacteria and fungi are the main contaminants of mammalian cell cultures. A culture contamination is usually observed by a sudden drop in pH (for some bacterial contaminants) and/or a cloudiness in the medium. If the culture becomes cloudy overnight and changes from red to yellow, it is likely to be contaminated. The colour change is associated with a decrease in pH which is indicated by the presence of phenol red in the medium. Under the microscope (at $\times 100$) some granulation may be observed between cells, although it is important to distinguish this from precipitates that may occur from the media, particularly in the presence of serum. At a magnification of $\times 400$ it may be possible to resolve individual bacteria, but unless the contamination is persistent it is not usually necessary to characterize it.

Contamination of cultures with mycoplasma may become a problem. Mycoplasmas derive from a genus of simple prokaryotes with a diameter of 0.2-2 μm. They have a restricted metabolism and are commonly associated with mammalian cells. They can be elusive because of their potential to pass through filters and they are not visible under low-power microscopy. Furthermore, the normal cocktail of penicillin and streptomycin used in cultures is ineffective because mycoplasmas do not have a cell wall. They are slow growing and consequently may not overgrow the cell culture. However, they may well affect the cellular growth rate, morphology, viability and metabolism.

Mycoplasma infects the cytoplasm of mammalian cells and may cause particularly recognizable effects on metabolism. For example, *Mycoplasma arginini* has a high requirement for arginine and will cause a rapid increase in culture pH. The best way to ensure a mycoplasma-free cell line is to test for mycoplasma contamination at regular intervals—every 3-6 months would be reasonable but this would depend upon usage. The test should be performed in an independent laboratory, so that positive controls could be used and the interpretation of results would be free of bias. Several commercial companies offer a service for mycoplasma testing. One type of test involves the use of fluorescent dyes specific for DNA such as Hoechst 33258 or DAPI (4',6-diamidino-2-phenylindole). Contaminated cells are identified by microscopic examination for the presence of fluorescent structures in the cytoplasm surrounding a stained cell nucleus.

If mycoplasma is detected, the best procedure is to autoclave the contaminated culture and decontaminate the culture laboratory. If the contaminated cells are irreplaceable elimination of the mycoplasma is possible by a cocktail of selected antibiotics, the composition of which depends upon the mycoplasma species. However, such a procedure of cell line decontamination is not always successful.

What to do if Cultures become Contaminated?

It is important that contaminated cultures are removed quickly from a culture laboratory before the contamination spreads to noninfected cultures. The cultures should be disinfected and autoclaved as soon as possible and without exposing the sterile area to an infected aerosol.

If stock cells are infected with mycoplasma then they also should be removed from the laboratory and destroyed. Techniques are available to eliminate mycoplasma contamination from cells by selective antibiotic treatment but the procedures are time consuming and only worth consideration if the cell stocks are irreplaceable.

The source of contamination is only worth investigation if the occurrence is frequent. Media contamination may be traced to faulty sterilization or storage in contaminated bottles. With inexperienced workers in cell culture the most probable source of contamination is from poor aseptic technique during culture manipulations.

Culture Conditions

Most cells in culture grow best at 37°C and at pH 7.4. If the cells are subjected to a temperature slightly lower than the 37°C optimum then the growth rate will be reduced but the cells should not be damaged. However, higher temperatures of 39-40°C will destroy the cells. So it is very important to be sure that the temperature does not increase in the incubator. The pH of cultures is maintained by a bicarbonate-CO_2 buffer system which is the main system that operates in blood *in vivo*.

$$CO_2 + H_2O \rightleftharpoons HCO_3^- + H^+$$

The pK_a of this buffering system is 6.3, which is adequate but not ideal for maintaining cultures at pH 7.4, which is the usual optimum for cell growth. The buffer equilibrium in the liquid phase is dependent on the presence of CO_2 in the gas phase. For pH control, an enriched CO_2 atmosphere is provided in the incubator.

A concentration of bicarbonate in the medium (normally 24 mM) maintains an equilibrium with CO_2 at a partial pressure in the gas phase of 40 mmHg (which corresponds to 5% CO_2). In some media formulations the bicarbonate concentration is higher (up to 48 mM). This increases the buffering of the media. However, the CO_2 in the atmosphere has to be set at 10% in order to maintain the pH of the cultures. DMEM cultures are grown in 10% CO_2. EMEM and RPMI cultures are grown in 5% CO_2.

Even with the bicarbonate-CO_2 buffer system the culture may still undergo pH changes during cell growth. In the early stages of culture a slight increase of pH may occur as a result of bicarbonate decomposition in the medium. However, as the cells grow and energy metabolism becomes well established, the production of lactic acid by cellular metabolism can cause a gradual decline in pH. This drift in culture pH during cell growth should not exceed a pH range of 6.8-7.6. The pH changes can be monitored by the colour of phenol red which is normally present in the culture medium. Colours of phenol red:

pH 6.5 = yellow; pH 7.0 = orange; pH 7.4 = red; pH 7.8 -= purple.

CULTURE MEDIUM

Cells are cultured in a chemically complex liquid medium suitable for supporting growth for several generations. There are many standard media formulations that have been developed for the growth of particular cell types.

To determine which media to use for a specific cell line, it is advisable to search the literature to see if previous experiments suggest the use of a particular media formulation. Alternatively, growth trials using three or four different media formulations may be performed. Some media, such as DMEM, have high concentrations of amino acids and vitamins and are suitable for prolonged cell growth. Other media, such as Ham's F-12, contain a wide range of different components which may be required to satisfy the fastidious requirements of some cell lines. Combinations of standard formulations can also be used for cell growth. For example, a 1:1 v/v mixture of DMEM and Ham's F-12 has been found to he effective as a good basal medium for serum-free formulations to support the growth of a number of cell lines.

Table 8.1. Commonly used culture media

Media	*Comments*
BME	Eagle's basal medium; originally designed for mouse L and HeLa cells
EMEM	Eagle's minimum essential medium; used for a wide variety of cell lines
DMEM	Dulbecco's modification of Eagle's medium; has *4X* the amino acid and vitamin concentration of BME
GMEM	Glasgow's modification of Eagle's medium; has 2x the amino acid and vitamin concentration of BME
RPMI 1640	Roswell Park Memorial Institute medium; used for lymphocyte and hybridoma cultures
Leibovitz	Used for fibroblast growth in the absence of a $C0_2$-enriched atmosphere
Ham's F- 12	Has a complex composition and used for a variety of cell lines
199	Extremely complex medium (61 components) and can support cell growth without serum should not be autoclaved or stored for any length of time. The half-life for glutamine in medium at 4°C is 3 weeks. Liquid medium can be purchased. free of glutamine or bicarbonate. In this condition it can be stored at 4°C until required.

Media can be supplied commercially as a 1× liquid ready for use or as a concentrate in liquid or powdered form. The concentrates are much cheaper and are advisable ii it is intended to use large volumes of one type of medium. The liquid concentrate (normally 10×) is provided sterile and requires dilution with presterilized distilled water. Powdered medium should be dissolved in water and sterilized by filtration through a 0.22 μm filter. Filtration can be mediated by suction with a vacuum pump or by liquid pressure from a peristaltic pump. The systems shown can be bought as presterilized disposable units. The water used for dilution has to be of high purity and should be prepared by reversed osmosis or double distillation. There are some unstable components (e.g. glutamine, bicarbonate) in culture medium and so complete liquid medium should not be autoclaved or stored for any length of time. The half-life for glutamine in medium at 4°Cis 3 weeks. Liquid medium can be purchased free of glutamine or bicarbonate. In this condition it can be stored at 4°C until required.

What are the Components of a Typical Culture Medium?

1. *Carbohydrates*. Glucose is used in most formulations to provide an energy source as well as a precursor for biosynthesis, such as ribose needed in nucleic acid synthesis. Lactic acid is the major product of glycolysis. In most cultures only a small proportion of glucose is completely oxidized via the tricarboxylic acid (TCA) cycle.
2. Alternative carbohydrates may be used, such as fructose, which results in a decreased lactic acid production and a more stable culture pit.
3. Amino acids are included at a concentration of 0.1-0.2 mM as a source of precursors for protein synthesis. Glutamine is normally included at higher concentrations (2-4 mM) in order to act as a

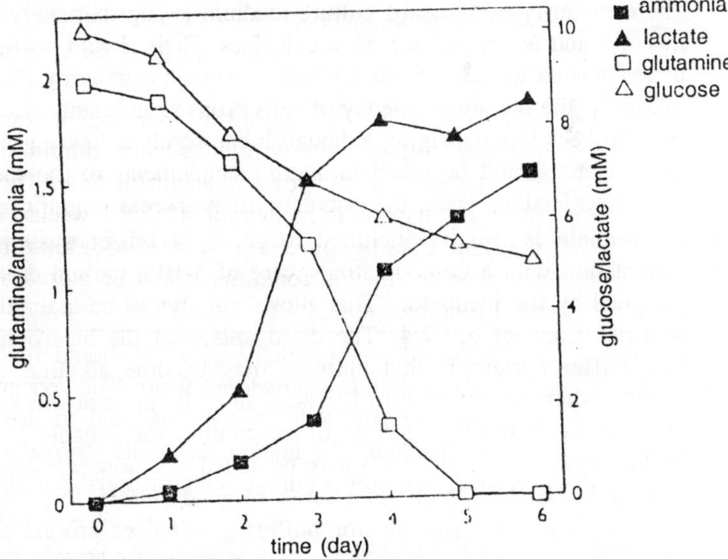

Fig. 8.3. Substrate utilization and by-product formation during the growth of a murine hybridization

precursor for the TCA cycle intermediates. However, ammonia is formed from the metabolic breakdown of glutamine and can be inhibitory to growth in some cultures. Figure 4.5 shows a typical pattern of substrate utilization and by-product formation in a hybridoma culture. The ammonia is produced from glutamine either by thermal degradation in the medium or by cellular metabolism.

4. Salts are included so that the solution is isotonic and has no imbalances with the intracellular content. Osmolarity is a measure of the total number of particles dissolved in a solution.

$$\text{Osmolarity} = \text{molarity} \times \frac{\text{moles of particles or ions}}{\text{moles of solute}}$$

Therefore, a solution at 1 Osmolar (or 1 Osm/l) has 6.023×10^{23} particles in 1 liter of water. Ionization increases the potential number of osmotically active particles. Therefore, 1 mM NaCl = 2 mOsm/l. Osmolality = mOsm/kg of water and is defined by a colligative property such as the ability to reduce the freezing point of water. Thus, a 1 Osm/kg solution reduces the freezing point of water by 0.001858°C. For practical purposes, there is very little difference between osmolarity and osmolality and both terms are used in the cell culture literature.

The osmolarity of standard culture medium is approximately 300 mOsm/l and is optimal for most cell lines. Cells should normally be maintained within 10% of this value. At high osmolarity (>400 mOsm/l), the plating efficiency of cells drops significantly. Lower osmolarity is less damaging although it can result in lower growth rates. Care should be taken in adding supplements to the media or in batchfeeding when, the osmolarity may increase significantly.

5. Bicarbonate is usually included to act as a buffer system in conjunction with a gaseous atmosphere of 5-10% carbon dioxide provided by the incubator. This allows cultures to be maintained at a pH range of 6.9-7.4. The disadvantage of the bicarbonate-CO_2 buffer system is that cultures may become alkaline very quickly when removed from the incubator. To prevent this or to increase the buffering capacity of the culture, the organic buffer HEPES (pK_a = 7.3 at 37°C) may be added at a concentration of 10-20 mM. HEPES is one of a series of good buffers, that are all organic compounds suitable for buffering at values around their pK. Other examples are MES (pK_a = 6.5), and CHES (pK_a = 9.5). HEPES is the optimal of these for cell culture and it has

the advantage that cultures can be grown without an enriched CO_2 atmosphere.

6. Vitamins and hormones are present at relatively low concentrations (μ molar quantities) and are utilized as metabolic co-factors. The content of vitamins and hormones varies enormously from one medium formulation to another. This reflects the variable requirement of cell lines for these components.
7. Phenol red is usually added as a pH indicator of the medium and accounts for the colour of culture media. This indicator is useful because it is particularly sensitive to slight pH changes around the growth optimum for cells. At lower pH the phenol red becomes orange (pH 7.0) or yellow (pH 6.5). An overnight change in colour of the culture from red to yellow usually indicates bacterial contamination.

Media Supplements

Serum

Serum is normally added to culture media at a concentration of 10% (v/v) to promote cell growth. Serum is the supernatant of clotted blood which contains undefined materials essential for cell proliferation.

Cow (bovine) or horse (equine) serum is most commonly used with fetal calf serum (FCS) being considered particularly effective because of its high content of embryonic growth factors. FCS will promote the growth of a wide range of cell lines, although the availability of this serum can vary. Other suitable bovine sources are 'newborn calf serum' that is processed from an animal under 10 clays of age or 'donor calf serum' that is processed from calves under 8 months. Donor horse serum is also widely used in cell culture.

Serum is purchased for laboratory use in 100 ml or 500 ml bottles from a cell culture supplier who will have treated the serum in a particular way. Typically serum is sterilized by filtration through at least one 0.1 μm filter. Each batch is then tested for a variety of microbial and chemical contaminants including bacteria, fungi and specific viruses depending upon the source of the serum. Special treatment of serum offered by suppliers may include the following.

1. Gamma irradiation involves exposure of the serum to radiation from a ^{60}Co source.
2. Heat inactivation involves incubation at 56°C for 30 minutes.
3. Gamma globulin removal involves ethanol fractionation which reduces the γ-globulin content of the serum. This is important for

monoclonal antibody production from hybridoma cultures because γ-globulin derived from serum could interfere with the extraction or assay of the antibody.

4. Dialysis usually involves serum dialysis against 0.15 M NaCl. This reduces the low-molecular-weight components of the serum but maintains its osmolarity. This treatment eliminates the amino acid, carbohydrate or nucleotide content of serum and might be required, for example, in a study of nutrient utilization of cells in culture.

FCS is the most widely used media supplement. However, it is expensive and often in short supply. The serum is generally tested for performance and cytotoxicity. Performance is measured by the ability of the serum to support plating efficiency and cell growth. The specifics of these tests are given by each manufacturer.

Are there Alternatives to Serum?

Although serum has been used extensively as a medium supplement to enhance cell growth, there are a number of widely recognized disadvantages connected with its use.

1. It is chemically undefined and variation between batches can result in inconsistent promotion of cell growth.
2. It is expensive. FCS can account for 70-80% of the cost of some formulations. This is an important consideration in large-scale cultures.
3. The proteins in serum can compromise the extraction and purification procedures for cell-secreted proteins.
4. Serum is vulnerable to contamination with infectious agents such as viruses and prions.

For these reasons considerable efforts have been made to develop supplements of hormones and growth factors to replace serum. These serum-free supplements are specific for particular cell types. Commonly used ingredients for these supplements include insulin, transferrin, ethanolamine and selenite. Preprepared serum-free media supplements are commercially available but although these are chemically defined, the manufacturers do not usually reveal the content. Any contaminants are more toxic to cells in low protein medium. So, serum-free formulations must be prepared in water of high purity.

For an extended study on a particular cell line, it is often worth spending some time developing a serum-free formulation based on a modification of a standard published recipe. When developing a serum-free formulation, it is very important to start with an optimal basal medium-rich formulations such as DMEM/Ham's F-12 have been found particularly suitable.

A gradual process of adaptation of cells from serum to serum-free medium may be necessary depending upon the cell line. To do this, cells are subcultured in successive media that contain reduced levels of serum. Eventually the medium is replaced by a serum-free formulation.

Table 8.2. A serum-free formulation used for growth of hybridomas

DMEM/Ham's F-12 (1:I vlv)	
Insulin	5 μg/ml
Transferrin	35 lag/ml
Ethanolamine	20 μM
Sodium selenite	1 nM

Antibiotics

Antibiotics are often included in media for short-term cultures in order to reduce the risk of contamination. The optimal concentration of antibiotics should be determined empirically bearing in mind that they may be cytotoxic. Antibiotics are often used in combination in culture medium and the following cocktail can be recommended for general use:

1. Penicillin G (100 U/ml) to inhibit the growth of Gram-positive bacteria;
2. Streptomycin (50 mg/l) to inhibit the growth of many Gram-negative and Gram-positive bacteria;
3. Amphotericin B (25 mg/l) as an antifungal agent.

However, the use of antibiotics for routine subculture or in stock cultures should be discouraged because low levels of bacterial or fungal contamination may be masked and may cause problems at a later date. Furthermore, extensive use of antibiotics may cause the selective retention of antibiotic resistant contaminants which can cause future problems.

Cell Metabolism during Culture

Glucose is normally present at an initial concentration of 10-25 mM in most media formulations and this decreases to around half the concentration level within the period of a batch culture. The major role of glucose is to provide the energy metabolism of the cells which occurs primarily anaerobically. The terminal product of glycolysis is lactic acid and this is reflected by an accumulation of this compound in the media to parallel the consumption of glucose. The molar ratio of lactate produced/glucose consumed may approach two and this is indicative of anaerobic glycolysis. The production of lactic acid may lower the pH of the culture and this may decrease the cellular growth rate. However, if the culture is appropriately buffered then the accumulated lactate that typically occurs in a batch culture should not cause any growth inhibition.

A secondary but very important function of glucose is to provide a substrate for the pentose phosphate pathway which provides the ribose necessary for nucleic acid synthesis. In fact, it has been shown in some cell cultures that glucose is not essential for cell growth providing there is an alternative source of ribose. Substitution of glucose by either fructose, galactose or maltose decreases the glycolytic rate and production of lactic acid. The metabolism of these alternative carbohydrate sources may proceed exclusively through the pentose phosphate pathway.

Next to the carbohydrate source, glutamine is the most abundant source of reduced carbon in the culture medium. Glutamine is normally included at a concentration of 1-5 mM, which is a significantly higher concentration than any of the other amino acids. Glutamine is an important precursor for the synthesis of purines, pyrimidines, amino sugars and asparagine. However, glutamine also has an important role as substrate for the TCA cycle and this is reflected in a high rate of utilization. The metabolic pathway for glutamine utilization (called *glutaminolysis*) follows two steps of nitrogen removal before conversion to 2-oxoglutarate (also known as *alpha keto glutarate*) which is an intermediate of the TCA cycle. Complete oxidation may occur although it is also likely that other products such as lactic acid or alanine may be by-products of this pathway.

The metabolic deamination of glutamine leads to ammonia which accumulates in the culture up to 2-4 mM and can be inhibitory to cell growth. The problem of ammonia accumulation is made worse by the fact that glutamine may decompose spontaneously to produce ammonia in the culture medium at a rate of 0.1 mM per day at 37°C. The extent of growth inhibition of ammonia is greater at higher pH and is cell line dependent. Substitution of an alternative carbon source such as glutamate may be possible and will result in a lower production of ammonia. However, cells require adaptation to enable an enhancement of the specific rate of membrane transport, which is normally low for glutamate. The accumulation of ammonia from glutamine can be decreased by a continuous feed of a low concentration of glutamine into the culture or alternatively by the use of glutamine-containing dipeptides that hydrolyze slowly in the culture.

The mixtures of defined amino acids are normally added to culture media as precursors for protein biosynthesis. However, the figure shows the high rate of glutamine utilization required as an energy source and the significant release of alanine into the medium by cells. The accumulation of alanine is a reflection of a mechanism of sequestration (or detoxification) of ammonia in the cell by amination of pyruvate, produced by glycolysis or by glutaminolysis. The accumulated alanine is not growth inhibitory and may in fact be utilized toward the end of the culture if the supply of other carbon sources decreases. Of the other amino acids, serine, aspartic acid and those with a branched chain are particularly well utilized. The sulfur-containing amino acids methionine and cysteine are essential and rapidly consumed.

PROTOCOL

Harvesting Anchorage-dependent Cells with Trypsin

Reagent

Trypsin reagent 4 ml trypsin (0.25%) + 16 ml versene in 0.2 g/l PBS

Method

1. Remove growth medium from culture flask by decanting or with a suction pump.
2. Wash cells in flask with PBS.
3. Add trypsin/versene reagent to completely cover the cell layer.
4. Incubate at 37°C for up to 15 min. Examine the cell layer at frequent intervals. Tap the flask gently to aid the dispersion of cells.
5. When the cell sheet is sufficiently dispersed, add growth medium (with serum) or trypsin inhibitor. Use approximately ×5 volume of trypsin solution left on the cells.
6. Transfer cell suspension into fresh medium. Dispense into new flasks - the subcultures.

Adaptation of Cells to a Serum-free Medium

Introduction

Some cell lines can be subcultured directly from a serum-based to a serum-free medium without loss of growth performance. However, in other cases cells may be adapted slowly into the new serum-free medium. The process of adaptation may involve changes in cellular metabolism or the induction of specific cellular growth factors.The following protocol is suitable in allowing such changes as cells are weaned to a new medium.

Method

1. Anchorage-dependent cells such as MDCK grow well in a medium such as DMEM supplemented with 5% calf serum. This can be used as a starting culture for adaptation to the serum-free medium (SFM).
2. Inoculate cells grown in the serum-based medium (SBM) into a medium containing 5:1 v/v SBM: SFM at a density $>2 \times 10^5$ cells/ml.
3. Subculture after 48 h growth for two passages.
4. Increase the content of serum-free medium by changing the SBM/SFM ratio to 4:1, 3:1.2:1.1:1,1:2,1:3, 1:5 and 1:10 before growing in 100% SFM.At each stage of medium dilution subculture (×2) as in step 3.
5. It may be necessary to repeat some of the steps of adaptation if cell growth is poor. This may occur at low levels of serum.

Trypsinization of Cells in Serum-tree Medium

Introduction

During the subculture of anchorage-dependent cells, the proteolytic activity of trypsin is used to detach cells from the surface of a culture flask. The exposure of MDCK cells to trypsin should be as short as possible (typically 20 min) to minimize the potential of proteolytic damage of the cells. In serum-supplemented cultures the addition of fresh medium is sufficient to neutralize the effect of the trypsin because of the high protein content. However, in low-protein serum-free media the trypsin must be neutralized by use of a specific inhibitor prior to subculture.

The method described below allows subculture with trypsin but without the requirement for serum.

Materials

Trypsin Inhibitor: 0.25% w/v soybean trypsin inhibitor (Gibco-BRL 17075-029) in Dulbecco's phosphate buffered saline (filter sterilized).

Method

1. Remove media from a 75 cm^2 T-flask and add 8 ml prewarmed PBS-EDTA.

2. Let the flask stand for 30 seconds then remove the PBS-EDTA
3. Add 1.5 ml prewarmed trypsin (0.25%). Incubate for 10-20 min at 37°C. Gently tilt the flask every few minutes to distribute the trypsin evenly.
4. The translucent monolayer should become more opaque. When this occurs tap the flask firmly 10 times against the palm of the hand to dislodge the monolayer of cells. If the cells do not dislodge continue to incubate them at 37°C.
5. Add 2.0 ml trypsin inhibitor 0.25%, swirl the flask and let stand for 1 minute.
6. Add 8 ml PBS-EDTA and break up the aggregates by pipetting up and down 10-20 times. Centrifuge at 300 g for 5 min.
7. Aspirate the supernatant and re-suspend the cells in an appropriate amount of prewarmed media—usually at 10^5 cells/ml to initiate a new culture.

Note

The quantity or concentration of trypsin addition may be varied depending on the cells and how firmly they are attached to the growth surface. It is recommended to use the same volume and concentration of trypsin inhibitor at stage 5 as trypsin used at stage 3.

9

GENE MANIPULATION IN CULTURE

The discovery of methods to isolate and transfer DNA from one species to another has led to numerous advances in understanding the function and regulatory mechanisms of individual genes. The technology has also been applied to genetically manipulate cells so that they become capable of expressing high levels of a selected protein ('overexpression'). This usually follows the transfer of a recombinant DNA incorporated into a vector that ensures rates of transcription and subsequent expression of the recombinant protein. The techniques could be the basis of designing experiments aimed at explaining the control of protein synthesis or cell differentiation. This is also the basis of large-scale production of the protein products of selected human genes. In this chapter, we will consider some of these techniques in relation to animal cell cultures. These possibilities have led to the increased commercial activity and interest in animal cell technology over the last 10 years.

NEED FOR MAMMALIAN CELLS

Mammalian cells have been the host cells of choice for the synthesis of complex glycoproteins from recombinant genes. They have the required characteristics for appropriate post-translational processing and product secretion. The most widely used host cell line for recombinant DNA has been the Chinese Hamster Ovary (CHO). The metabolism of this cell has been well studied and characterized. A number of vectors have been designed to provide high yields of a range of biologically active mammalian proteins from these cells.

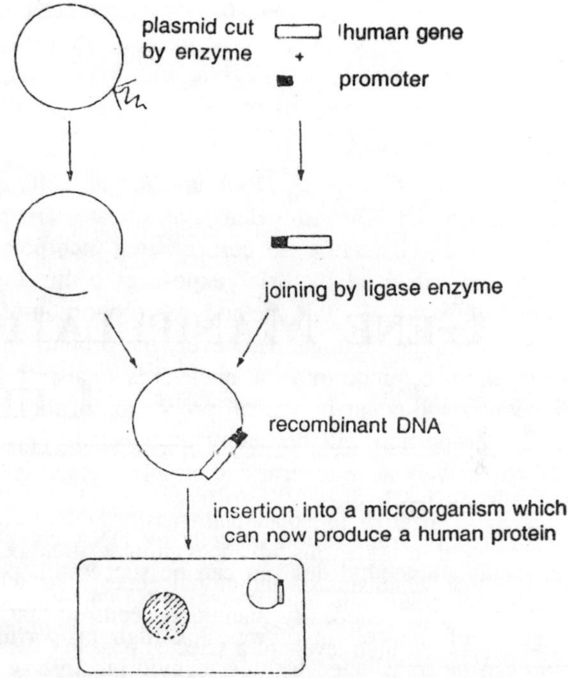

Fig. 9.1. Transfection of a human gene into a cell with a plasmid.

How to Get DNA into Mammalian Cells

Transfection means the transfer of nucleic acid into mammalian cells although it can also be referred to as DNA-mediated transformation. This should not be confused with the alternative use of 'transformation' meaning a change of growth characteristics of animal cells from a finite to an indefinite lifespan. The ability of mammalian cells to incorporate foreign DNA was first shown in the early 1960s when mutant human cells which lacked the enzyme hypoxanthine-guanine phosphoribosyl transferase (HGPRT) were transfected with isolated human nuclear DNA. Transfected cells were subsequently found with a functional gene for the HGPRT enzyme.

A variety of methods is now available for incorporating foreign DNA into animal cells. However, this does not always lead to the expression of a corresponding protein. Protein expression is dependent upon the incorporation of a structural gene along with the required expression signals into the cell.

Here, we will consider some of the methods that can be used to introduce the DNA into the cells. One important criterion for choosing a particular method is that the yield of transfected cells is sufficiently high to allow easy isolation. Also, some methods are better than others in allowing the foreign DNA to integrate with the cell's nuclear DNA and allowing stable expression of protein.

Transfer of 'Naked' DNA

Calcium phosphate method. DNA uptake into cells can be aided by coprecipitation of DNA with calcium phosphate. The precipitate is adsorbed onto the cell membrane before being incorporated into the cells, a process enhanced by brief exposure to dimethyl sulfoxide (DMSO) or to glycerol. By this method, the proportion of cells taking up the DNA precipitate is high. However, the proportion of the cells that take up the precipitate into the nucleus is relatively low (1-2%). Gene expression will occur in a small proportion of these cells and an even smaller proportion will be stably transfected. This method has been used extensively as a general and simple method for introducing DNA into mammalian cells.

DEAE-dextran. Transfection of cells by DNA complexed to the polymer, diethylaminoethyl dextran can be successful, particularly if accompanied by treatment with DMSO, which causes temporary destabilization of the cell membrane. A high proportion of a cell population can be transfected by this method but this generally leads only to transient protein expression.

Encapsulation of DNA in liposome vesicles is an effective way to improve the efficiency of cellular uptake. The liposomes are complexes formed from a positively charged lipid and DNA. The complexes are readily taken up by mammalian cells. A suitable cationic lipid reagent is Lipofectamine, which is a proprietary formulation supplied by Invitrogen. The reagent is complexed with DNA or RNA prior to transfection of a suspension of preconfluent cells. The procedure allows rapid high-efficiency transfections that result in high levels of gene expression.

Electroporation involves the exposure of a cell suspension to a highvoltage electrical impulse. Transient pores that are formed in the cell membrane promote the uptake of DNA. The conditions for electroporation are determined experimentally so as to maximize the number of surviving transfected cells. Typically, the cells are subjected to electrical impulses of 100-200 V for 1-2 ms.

Micro-injection involves the insertion of DNA by a micro-needle into the nucleus of individual cells. This is an effective technique for inserting DNA into a small number of cells but is generally only performed by a specialized laboratory. 'Hits' are almost certain but the technique cannot be applied to a large number of cells.

Protoplast fusion. Plasmids which are grown in bacteria can be introduced directly into animal cells by cell fusion with the bacterial protoplast. The protoplast is prepared by lysozyme treatment of the bacteria to remove the cell wall. The bacterial protoplast is fused to a mammalian cell by treatment with polyethylene glycol. The entire content of the bacterial cell is introduced into the mammalian cell and this has the advantage of not requiring DNA purification. I owever, difficulties can arise from the introduction of unwanted bacterial components into the recipient cells.

DNA Transfer using Viruses

Viruses are efficient vectors for DNA transfer—a process termed transduction. Animal viruses have important characteristics that increase the efficiency of nucleic acid transfer into mammalian cells:

1. They promote the transfer of DNA to cells because of capsid proteins that bind to cell membrane receptors;
2. They contain promoters that allow expression of inserted genes in animal cells;
3. They replicate to high copy numbers, that is produce several copies of the required gene within each cell;
4. Some viruses integrate efficiently into the animal cell genome.

The most widely used viral expression system has been derived from a simian virus, SV40. The virus contains a circular DNA duplex of 5.2 kilobases and many well-characterized sites of cleavage by restriction endonucleases. During SV40 infection of mammalian cells there are two phases of transcription. In the early phase, proteins known as the 'T antigens' are synthesized. This causes unwinding of the viral DNA leading to the latephase transcription and production of the viral capsid proteins. The infectious cycle takes about 70 hours and eventually leads to cell lysis.

In recombinant virus vectors, part of the viral genome is replaced by foreign DNA. These can be constructed as early or late region replacement vectors. The advantage of this type of vector is that DNA is introduced into the cell by viral infection—a process which is extremely efficient. However, the system has a limited host range (only monkey cells) and is suitable only for small foreign DNA inserts, The potential size of the foreign DNA insert can be increased if only a small portion of the SV40 DNA is utilized, such as the promoter region. This has been used extensively as a transient expression system.

Efficient vectors for integration of foreign genes into animal cells have been developed from retroviruses. Retroviruses can integrate genes into the nuclear DNA. These viruses have an RNA genome which requires the formation of complementary DNA copies by the enzyme reverse transcriptase. In the normal process of infection, the DNA can be integrated into the host cell's genome without causing cell lysis. The retroviruses can be used for cell immortalization.

Recombinant vectors are constructed by replacement of the retroviral structural genes with foreign DNA. They have the ability to replicate and integrate into the host cell's DNA but not for forming viral coat proteins.

Table 9.1. Advantages and disadvantages of methods of introducing DNA into cultured animal cells

Method	*Advantage*	*Disadvantage*
Calcium phosphate	Simple	Low incorporation into nucleus
DEAE-dextran	Simple	Transient expression only
Liposomes	Simple	Relatively unproven
Electroporation	Good for nonadherent cells	No cotransfection
Micro-injection	Efficient	Efficient/technically difficult
Protoplast fusion	Good for nonadherent cells	Variable results
SV40 virus infection	Efficient	Cell-type restricted
Retrovirus infection	Efficient	Cell-type restricted

How to Select and Amplify the Genes of Transfected Cells

Cell Selection by Genetic Markers

The proportion of cells that accept exogenous DNA during a transfection process may be as low as 1 in 10^6 and therefore a means of selection of the transfected cells is essential. To allow selection, a marker gene can be incorporated into a recipient cell. The purpose of the genetic marker is to identify the cells that have been transfected. The marker can be linked to the same DNA vector as the target gene. However, direct ligation of the two genes may not be necessary because of the phenomenon of cotransfection. This follows the observation that more than one gene can be transfected into a cell at the same time even if they are located on separate plasmids. This is explained by the fact that although only a few cells take up exogenous

DNA, the recipient cells accept relatively large amounts of DNA. Gene markers that have been used extensively in mammalian cells include genes for the enzymes hypoxanthine guanine phosphoribosyl transferase (HGPRT), thymidine kinase (TK) and dihydrofolate reductase (DHFR). These are nondominant marker genes and can only be used for selection in a population of cells that are mutants, lacking the marker enzyme activity prior to transfection.

Dominant genes are more useful because they do not require mutant recipient cells and can be used with any cell type. One example is the gene for the enzyme xanthine-guanine phosphoribosyl transferase (XGPRT) which is only present in bacteria and permits the metabolism of xanthine. Only recipient mammalian cells that have acquired the bacterial gene during transfection can survive in a medium containing xanthine, hypoxanthine and mycophenolic acid, which inhibits the conversion of IMP to XMP. Selection can be improved by the addition of aminopterin which inhibits *de novo* nucleotide synthesis, staking the cells entirely dependent upon the bacterial enzyme for purine nucleotide synthesis. Antibiotic resistance genes are also suitable as dominant markers. Fcr example, the neomycin-resistance gene (neo^R) codes for the enzyme aminoglycoside phosphotransferase. The presence of this enzyme confers resistance to the aminoglycoside group of antibiotics of which G418 is the one most commonly used. G418 prevents replication of mammalian cells. If the neo^R gene is expressed then the cells survive.

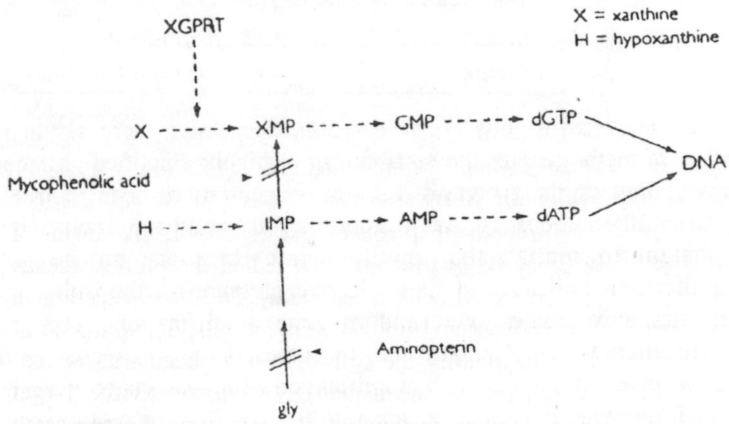

Fig. 9.2. The effect of the XGPRT selection system on purine metabolism.

Gene Amplification

'This involves an increase in the number of copies of a specific gene per cell and can result in increased synthesis of the corresponding protein. Gene amplification was first demonstrated in mammalian cells by selection for resistance to methotrexate which is an analog of folic acid and competitive inhibitor of the dihydrofolate reductase enzyme, DHFR. This enzyme is essential for the conversion of the folate co-enzyme required in purine and pyrimidine metabolism. Cellular resistance to the effects of methotrexate can occur by a stepwise increase in the concentration of methotrexate in the culture medium. Cells that become resistant to methotrexate fall into one of three classes: (a) reduced methotrexate uptake; (b) point mutation in the dhfr gene making the enzyme less susceptible to inhibition; and (c) multiple copies of the dhfr locus allowing the production of sufficient enzyme to compete with the effect of the inhibitor. In the third class, resistance is brought about by the amplification of the gene for dihydrofolate reductase (DHFR).

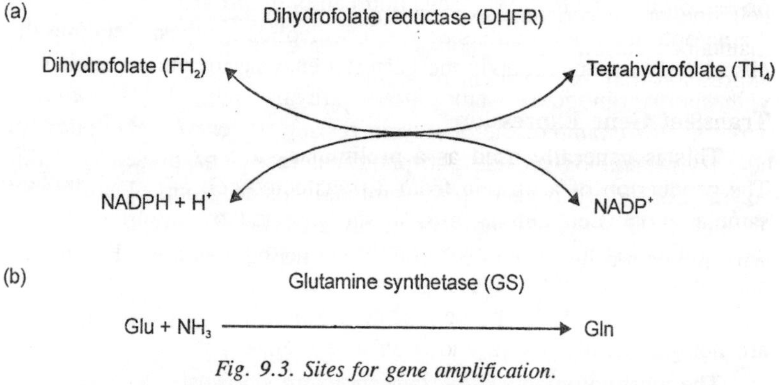

Fig. 9.3. Sites for gene amplification.

These resistant cells are selected by successive increases in the dosage of methotrexate. the cells can have highly amplified dhfr gene arrays, allowing the growth of cells in concentrations of inethotrexate up to $\times 10^5$ the dose lethal to nonresistant wild-type cells. It is important to realize that methotrexate does not induce gene amplification, rather it provides selective pressure for the isolation of cells that have undergone random gene amplification. The gene amplification is stable and results in corresponding increases in the enzyme levels in the cell. Co-amplification of an associated gene will occur if the gene is inserted adjacent to the DHFR gene on the vector.

By this method stable cell lines can be produced with as much as a 1000-fold increase in gene copy number and corresponding protein expression. An alternative system involves the gene for glutamine synthetase which is amplified by addition of methionine sulfoximine. The enzyme is required for the conversion of glutamate to glutamine which is essential for the growth of cells in the absence of glutamine.

How to Get Genes to Express Proteins

The insertion of DNA into a cell does not guarantee gene expression and synthesis of the corresponding protein. DNA transcription requires a number of expression signals which are specific for mammalian cells. The expression signals are usually added to the gene before it is transfected into a cell and is referred to as an 'expression system'. Various ways have been devised to increase the chances of the expression of proteins by isolated mammalian genes. The efficiencies of expression vary and depend upon the gene, expression signals and the cell line used. Stable gene expression is dependent upon the integration of a segment of DNA into the nucleus of the host cell line or an efficient extrachromosomal replication system that is maintained during cell division. Depending upon the method used, gene expression may be transient or stable.

Transient Gene Expression

This is generally used as a preliminary test of gene expression. The production of a protein from a transfected cell is assayed from a sample of the cell culture after a short period of time—usually 1-3 days following the uptake of DNA. For this, the transfected cells need not be selected or isolated but the rate of incorporation and expression of DNA in the cells needs to be high. However, the cells are not genetically stable and soon lose their expression ability.

The most commonly used transient expression system involves COS cells. These are monkey kidney cells transfected with a mutant SV40 virus having an inactivated origin of DNA replication (six basepair deletion). The SV40 DNA is not replicated because of this deletion. However, any transfected DNA containing a functional SV40 origin of replication can be replicated in these cells. A suitable vector contains the small SV40 DNA fragment as well as any cloned gene of interest. These can be transfected into COS cells to obtain high copy numbers of the cloned gene (2-5 $\times$ 10^5 copies/cell). However, permanent cell lines can not be established by this method because the extensive viral replication eventually has an adverse effect on the viability of the cells.

Stable Gene Expression

Stable expression systems are used to construct and isolate transfected cells that are genetically stable and are capable of indefinite synthesis of a particular protein. Recombinant vectors can be constructed that maximize the chances of obtaining such cells. The limiting factor in obtaining stable transfected cells is usually the frequency of DNA integration, rather than the frequency of cellular uptake. The ideal recombinant cell is one that grows with a doubling time of ~18 h, is genetically stable and secretes a recombinant protein at ~20°% of total cellular protein.

The most commonly used cell line for the stable expression of a recombinant protein is the Chinese hamster ovary (CH0) line—primarily because the cells have been well studied and characterized. Vectors are constructed with a variety of domains to ensure efficient, high-level expression of the desired protein in mammalian cells. Many of these vectors have nucleotide sequences from the bacteria plasmid pBR322. This enables vector replication in bacteria or mammalian cells (termed a *shuttle vector*).

Expression Cassettes

A number of expression signals are needed to ensure that a gene is expressed in mammalian cells. These signals are common to a wide variety of genes and can be used in a range of cell lines. An expression cassette is a DNA sequence containing these signal sequences that is incorporated into a vector used to transfect mammalian cells. The signal sequences of an expression cassette include the following.

1. A promoter that is required to initiate transcription of the gene. This marks the point at which transcription of a gene starts following RNA polymerase binding. Promoters contain highly conserved sequences (for example, the 'TATA' box or 'initiator') located upstream of the initial site of transcription. Promoters taken from an animal virus (for example, SV40, adeno- or retroviruses) have been used to express a number of transfected genes. These are described as 'strong promoters' because they allow a high rate of transcription.
2. A ribosome-binding site is a short nucleotide sequence required for attachment of mRNA to ribosomes.
3. A terminator is required downstream of the expressed gene to mark the point at which transcription stops.

Figure shows the elements of a shuttle vector which contains an expression cassette. This recombinant vector, named SVX, contains the following elements:

(i) Promoter and termination control regions from) a retrovirus;

(ii) The mammalian gene inserted for expression—in this example globin cDNA is inserted;

(iii) A selectable marker gene for neomycin(neo) resistance;

(iv) Origin of replication sites (ori) enabling replication in either bacteria or mammalian cells.

REGULATION OF GENE EXPRESSION

The ability to regulate the expression of a foreign gene by the addition of an external stimulus can have distinct advantages. In some cases, overexpression of a specific protein can be growth inhibitory or toxic to cells. So, it would be an advantage to switch off gene expression during cell growth and only allow the foreign gene to he expressed when maximum cell density is attained. The type of switching mechanism requires a regulated promoter such as the one isolated from mouse mammary tumour virus, which can be induced by dexamethasone. The inducible promoter can be used to induce the transcription of any gene associated with it in an expression vector when dexamethasone is added to the culture medium. The presence of the promoter can also enhance the level of protein expression.

PROTOCOL

Transfection of CHO-KI Cells by Lipofection

Materials

OptiMEM 1 (Invitrogen), Lipofectamine 2000 (Invitrogen).

Method

1. Prepare cells. Incubate 1.5×10^5 cells in 0.5 ml growth medium for 24 h in each well of a 24-well plate so that the culture becomes >90% confluent.
2. Prepare the DNA-lipid complex
   - Dilute 0.8 μg DNA with 50 μl OptiMEM.
   - Dilute 2.5 μl Lipofectamine 2000 with 50 μl OptiMEM and incubate for 5 min at room temperature.
   - Combine the diluted DNA and Lipofectamine. Incubate at room temperature for 20 min to form the DNA-lipid complex.
3. Transfection
   - Add the DNA-Lipofectamine complex (100 μl) to each well of cells.
   - Mix gently by rocking the plate and incubate at 37°C for at least 24 h before testing for transgene expression.

Transfection by Electroporation

Method

1. Suspend cells in serum-free medium.
2. Add 10 μg DNA or RNA to 10^7 cells.
3. Electroporate at 688 V/cm; 800 μF using the Cell-Porator Electroporation System (Gibco BRL).

Development and Isolation of Methotrexate-resistant CHO Cells

Materials

CHO dhfr cell culture in exponential growth.
Methotrexate in 0. 15 M NaCl (2.5 mg/ml).

Method

1. Inoculate the CHO cells from an exponentially growing culture into replicate T-flasks containing 0, 0.01, 0.02, 0.05 and 0.1 μg/ml methotrexate added to the culture media from the stock solution.
2. Incubate for 5 to 7 days.
3. Replace with medium containing the same amounts of methotrexate and allow to grow for another 5 to 7 days.
4. When the cell density in each flask reaches 2-10 $\times$ 10^6 cells/flask, remove the cells and inoculate into medium containing the same and higher ($\times$2 and $\times$10) concentrations of methotrexate.
5. Allow to grow for 5-7 days and select viable cells for subculture. Isolate small proportions of cells showing clonal growth in a background of enlarged or dying cells.
6. Continue selection and subculture with progressively higher methotrexate concentrations until the desired level of resistance is obtained. This can take 3-6 months for high levels of resistance.

Notes

It is important to store cell samples from subcultures periodically. This will safeguard against loss of cultures by poor growth or contamination. If cell samples are stored, then the procedure can be resumed at the previous point of methotrexate resistance.

10

PROTEINS IN CELL CULTURE

Most large proteins derived from animals have attached carbohydrate structures (glycans) that are added as a post-translational modification. At one time these glycans were thought to be unimportant to glycoprotein activity. However, now it is clear that an understanding of the carbohydrate component (glycan) of a recombinant glycoprotein is of importance for two main reasons. Firstly, the carbohydrate structures attached to a protein can affect many of its properties. This includes its pharmacokinetics, bioactivity, secretion, *in vivo* clearance, solubility, recognition and antigenicity, all of which influence the overall therapeutic profile of the glycoprotein. Secondly, quantitative and qualitative aspects of glycosylation can be affected by the production process in culture, including the host cell line, method of culture, extracellular environment and the protein itself.

Glycosylation is a process that occurs in eukaryotic cells in which oligosaccharides are added to the protein during synthesis and processed through the endoplasmic reticulum (ER) and Golgi apparatus along the secretory pathway. Glycoproteins occur as heterogeneous populations of molecules, called *glycoforms*. Glycomics is a term used for the analysis of heterogeneous pools of glycans. It is important to realize that the structure of the glycan is not governed by a predetermined template as is the case for proteins and a variety of structures are possible for any protein. The potential variability of glycoforms presents a difficulty to industrial production and for regulatory approval of therapeutic glycoproteins. The challenge is to know how the glycoprotein heterogeneity is generated, and how to evaluate its significance with respect to the safety and efficacy of the product.

The glycoforms have characteristic profiles that can vary with the control parameters of bioprocessing. It is important to understand these culture control parameters to assure the reproducibility of a bioprocess to yield the same glycoform profile that went through clinical trials and to maximize the titer of the desired glycoforms.

There are several advantages in using mammalian cells for the production of recombinant glycoproteins. One of the most important of these is their ability to perform post-translational modifications and achieve a product close to that produced *in vivo*. The use of other expression systems such as yeast, plant or insect cells is more limited because despite their potential economic advantages for culture and the higher yields of these systems, their glycosylation capacities do not resemble those of the mammalian cells.

Glycosylation engineering is gaining importance as a tool to optimize desirable properties such as stability, antigenicity and bioactivity of glycosylated therapeutic pharmaceuticals. This is achieved by genetic engineering of the pathways of oligosaccharide synthesis in the mammalian host cells. Incorporation of new glycosyltransferase activities can modify the product, compete with endogenous enzymes to produce novel glycoforms or maximize the proportion of beneficial ones.

The purpose of this chapter is to outline the capabilities of various cell lines for recombinant glycoprotein production. Each cell line's capacity for glycosylation may be modified by parameters associated with growth of the cells in culture. These combined factors will lead to a heterogeneity of glycoforms that will be considered in relation to their acceptability for use as human therapeutic agents.

Glycan Structures Present in Glycoproteins

There are three types of glycan structures attached to proteins: N-glycans, O-glycans and GPI anchors.

N-glycans

This is the most prevalent and widely studied form of glycosylation. The glycan is bound via an N-glycosidic bond to an Asn residue within the consensus amino acid sequence (sequon) Asn-X-Ser/Thr. However, the presence of this sequon in a protein does not guarantee glycosylation. The glycosylation of the sequon is variable and gives rise to a macroheterogeneity of glycoforms (variable site occupancy). Amongst other factors, the site occupancy will depend upon the tertiary structure of the protein. There may be a multiplicity of glycan structures at a particular site. Differences in these structures is referred to as

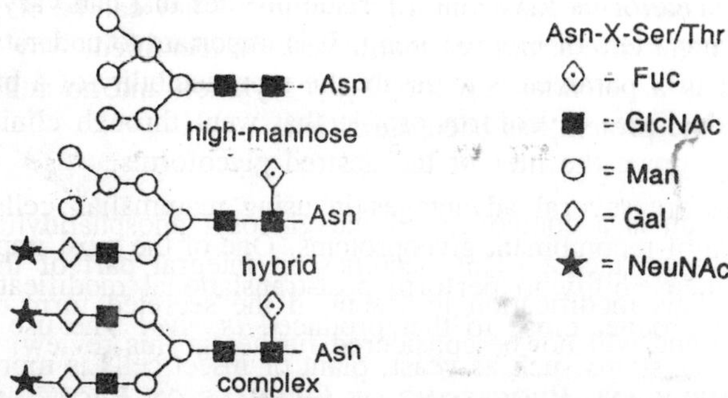

Fig. 10.1. N-glycan structures of glycoproteins.

microheterogeneity. The structures of N-linked glycans fall into three main categories: high-mannose, hybrid and complex type. They all have the same core structure, $Man_3GlcNAc_2$-Asn, but differ in their outer branches.

1. High-Mannose type: typically has two to six additional Man residues linked to the core.
2. Complex type: contains two or more outer branches containing N-acetyl glucosamine (GlcNAc), galactose (Gal), and sialic acid (SA).
3. Hybrid type: has features of both high-Man and complex type glycans.

Common substituents to the N-glycan structures are Fucose (Fuc) linked to either the innermost core GlcNAc (proximal) or the outer arm GlcNAc (peripheral). Also, a 'bisecting' GlcNAc may be linked by (β1,4 to the central core Man residue. The advantage of this method is that the type of linkage between monosaccharides can be shown.

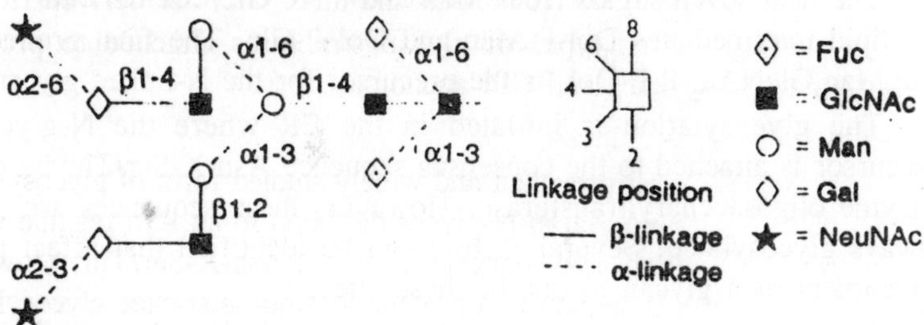

Fig. 10.2. Di-sialylated biantennary structure showing linkage positions.

O-glycans

The glycan is bound via an O-glycosidic bond to a Ser/Thr (O-glycosylation). The most common are the O-linked mucin-type glycans, which are a heterogeneous group of structures. They can be classified into one of eight core structures that have been identified. Any Ser or Thr residue is a potential site for O-glycosylation and no consensus sequence in the protein has been identified.

GPI anchor

The glycan is a component of the glycosyl phosphatidylinositol (GPI) membrane anchor. This becomes an integral part of the cell membrane. This modification is absent in the secreted form of any glycoprotein and will not be considered further in this review.

Assembly and Processing of Glycans on Proteins

Assembly of N-linked Glycans

This process is initiated in the endoplasmic reticulum as a protein is being synthesized from its constituent amino acids. The precursor glycan structure is added to the N-glycan sites and this is followed by a series of trimming reactions. High-mannose-type structures are formed during the processing in the ER. However, the completion of the processing to form hybrid and complex glycans takes place in the Golgi. The full diversity of glycan structures occurs because of a series of competing processing enzymes associated with this pathway.

All N-linked glycans on a glycoprotein share the same core because they all come from the same precursor which is transferred to the nascent protein in the endoplasmic reticulum. The precursor consists of a lipid (dolichol) linked to a glycan by a pyrophosphate bond. During synthesis of this glycan, sugars are added in a step-wise fashion, the first seven sugars (two GlcNAc and five Man) derive from the nucleotide sugars UDP-GlcNAc and GDP-Man respectively.

The next seven sugars (four Man and three Glc) are derived from the lipid intermediates Dol-P-Man and Dol-P-Glc. The final product, $Glc_3Man_9GlcNAc_2$-P-P-Dol, is the precursor for the N-linked glycans.

The glycosylation is initiated in the ER where the N-glycan precursor is attached to the consensus sequence Asn-X-Ser/Thr by the enzyme oligosaccharyltransferase. However, these sequences are not always glycosylated. Several factors can be identified that affect the attachment of a glycan to one of these sites.

1. The spatial arrangements of the peptide during the translation process can expose or hide the tripeptide sequence.
2. The amino acid sequence around the attachment site (Asn-X-Serf/Thr) is an important determinant of glycosylation efficiency. X can be any amino acid except Pro or Asp. The occupancy level is

high when X= Ser, Phe, intermediate for Leu, Glu and very low for Asp, Trp and Pro.

3. The availability of precursors (lipid, nucleotide sugars and correctly assembled precursor) level of expression of the oligosaccharyltransferase enzyme disulfide bond formation, which can make the site inaccessible to the precursor.

Glycan Processing

Processing is initiated by a series of glucosidases, which air located in the membrane of the endoplasmic reticulum. The newly synthesized glycoproteins are then transported to the Golgi çisternae by means of vesicles. In the Golgi a series of glycosidases and glycosyltransferases act on the N-linked glycans and lead to a diversity of structures (microheterogeneity). The final products can be modified further in the Golgi by the addition of:

1. Sialic acid: two main enzymes compete for the terminal Gal: α2,3 sialyltransferase (α2,3ST) and α2,6 sialyltransferase (α2,6ST);
2. Poly-N-acetyl lactosamine: it is added by βN-acetylglycosaminyltransferase (βGnT);
3. Fucose: by α1,6 fucosyltransferase, that adds fucose in an α1,6 linkage to the GlcNAc attached to Asn, or by α1,3 fucosyltransferase that adds fucose to C3 of the GlcNAc residue in the sequence Gal β1-4 GlcNAc β1-2 Man. The product is not a substrate for the α2,6ST. Fucose can be added at any time after M5 is synthesized but not after the action of Gal T or GnT III. Fucosylated glycans, however, can be modified by these enzymes.

The variety of glycans produced is a result of competitive activities of this range of enzymes. All the transferase-catalyzed reactions use sugar nucleotide co-substrates. It is evident that the key factor in determining the synthesis of particular N-linked glycans is the level of expression of the different glycosyltransferases.

The glycan profiles of glycoproteins are normally characteristic of the cell in which the protein is expressed and depends on cellular factors such as:

- Enzyme repertoire;
- Competition between different enzymes for one substrate;
- Transit time of the glycoproteins;
- Levels of sugar nucleotide donors;
- Competition between different glycosylation sites on the protein for the same pool of enzymes.

At any time, many glycoproteins may be trafficking through the glycosylation pathway, competing for the glycosylation enzymes. The glycans attached to the glycoprotein are processed by some enzymes and not by others. Umaha and Bailey (1997) proposed a mathematical model based on the activities of a set of eight enzymes and 32 reactions to determine the distribution of glycans into the major structural classes: high Man, hybrid, bisected hybrid, bi-, tri-, tri and tetraantennary complex and bisected complex glycans so the proportion of these structures could be calculated based on the kinetics of these enzymes.

Assembly of O-linked Glycans

O-linked glycans are added post-translationally to the fully folded protein. Glycosylation can occur on exposed Ser or Thr residues but no consensus sequence has been identified.

The most commonly found O-glycans are the mucin-type, although other structures such as O-linked fucose or O-linked glucose do exist. The first step for the assembly of the mucin-type O-glycans is the addition of an N-acetylgalactosamine (GalNAc) residue to a Ser/Thr by a GalNAc transferase (GalNAcT) from UDP-GalNAc.

Although no consensus sequence has been identified for O-glycosylation, the glycosylated residue is often associated with regions of the peptide that contain a high proportion of Ser, Thr and Pro. It appears that this would make the polypeptide assume a favourable conformation making the Ser or Thr more accessible. Thr residues appear to be glycosylated more efficiently than Ser. Various GalNAcT have been identified, having a broad but different substrate specificity and are expressed in a tissue-specific manner. Further elongation leads to a large number of structures, synthesized by various glycosyl-transferases, producing the eight different core structures. These core structures can be further modified by sialylation, fucosylation, sulfation, methylation or acetylation. A characteristic of transformed cells is that the initial GalNAc residue is not elongated and is known as the Tn antigen. Each of the core structures can give rise to a series of modified structures.

Glycoprotein Analysis

Considerable advances have been made in the analysis of N-glycan structures over the last few years. These glycans can be separated from proteins by a specific enzyme (N-PNGase F), a cleavage that can take place in isolated bands from a protein gel. The isolated glycans may then be labeled with a fluorescent dye and systematically

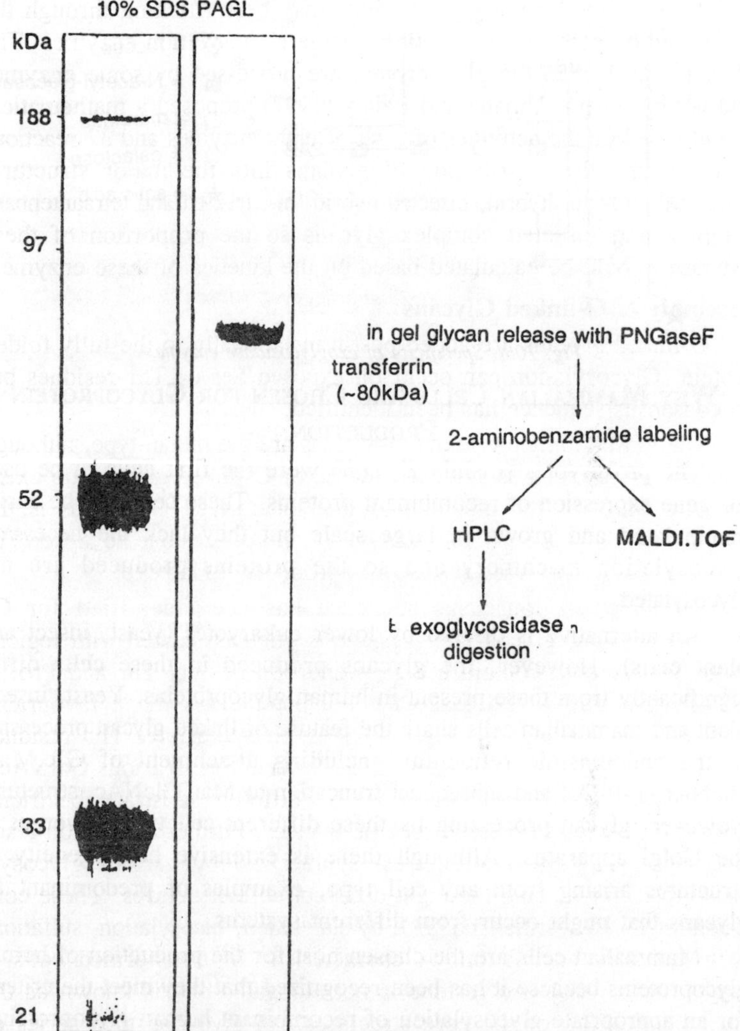

Fig. 10.3. Glycoprotein analysis.

degraded by specific exoglycosidases to give data related to the sequence of the glycan. Sequential addition of the enzymes results in a step-wise reduction in the size of the glycan and these products can be identified by HPLC. A simple method of glycan separation can also be provided by gel electrophoresis, providing the glycans are charged. The fluorophore, ANTS offers a suitable glycan label which provides both a negative charge and a fluorescent group for detection.

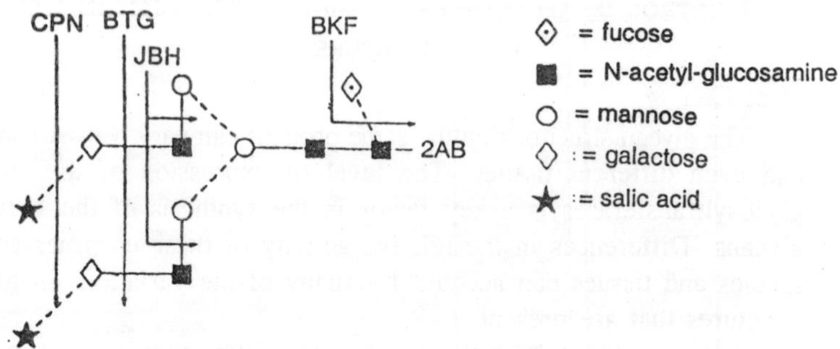

Fig. 10.4. Specificity of exoglycosidase enzyme.

Why Mammalian Cells are Chosen for Glycoprotein Production

The prokaryotes (mainly *E. coli*) were the first cells to be used for gene expression of recombinant proteins. These cells can be easily manipulated and grown in large scale but they lack the necessary glycosylation machinery and so the proteins produced are not glycosylated.

An alternative is offered by lower eukaryotes (yeast, insect and plant cells). However, the glycans produced in these cells differ significantly from those present in human glycoproteins. Yeast, insect, plant and mammalian cells share the feature of linked glycan processing in the endoplasmic reticulum, including attachment of GIc_3Man_9 $GlcNAc_2$-P-P-Dol and subsequent truncation to $Man_8GlcNAc_2$ structure. However, glycan processing by these different cell types diverges in the Golgi apparatus. Although there is extensive heterogeneity of structures arising from any cell type, examples of predominant N-glycans that might occur from different systems.

Mammalian cells are the chosen host for the production of human glycoproteins because it has been recognized that they meet the criteria for an appropriate glycosylation of recombinant human glycoproteins. They are capable of complex-type N-glycan processing whereas the other systems are not. However, there are different capabilities for glycosylation between mammalian cell lines. The most commonly used cell lines for recombinant protein production are the hamster-derived *Chinese hamster ovary* (CHO) and *baby hamster kidney* (BHK) cells. These cell lines have been chosen because of their favourable growth characteristics in culture as anchorage-dependent or suspension cells. They have been used as expression systems to produce proteins whose glycoforms are similar to the native human products.

Control of Glycan Processing in Mammalian Cell Culture

Host Cell

The glycan structures on the same proteins can vary between species and even different tissues. The level of expression of the various glycosyltransferases is a key factor in the synthesis of the N-linked glycans. Differences in the relative activity of these enzymes among species and tissues can account for many of the variations in glycan structures that are present.

An analysis of the glycan structures of IgG from 13 different species shows that there is significant variation in the proportion of terminal galactose, core fucose and bisecting GlcNAc. The predominant monosaccharide precursors may also differ structurally between species. For example the terminal sialic acid found in glycoproteins from goats, sheep and cows is predominantly N-glycolyl-neuraminic acid (NGNA) rather than N-acetyl-neuraminic acid (NANA) which is the sialic acid structure generally found in humans and rodents.

CHO and BHK are the most commonly used cell lines for the production of recombinant proteins with potential application as therapeutic agents. For such an application it would be desirable to obtain proteins with as near a human glycosylation profile as possible. However both CHO and BHK show differences in their potential for glycosylation compared to human cells. The sialyltransferase enzyme; α2,6ST is not active in these cell lines, leading to exclusively α2,3-linked terminal SA residues. Furthermore, the absence of a functional α1,3 fucosyltransferase in CHO cells prevents the addition of peripheral Fuc residues and also the absence of N-acetylglucosaminyltransferase III (GnT III) prevents the addition of bisecting GlcNAc.

However, the differences in glycosylation capabilities between CHO and human cells do not appear to result in glycoproteins that are immunogenic. Natural human erythropoietin (EPO) consists of a mixture of sialylated forms—60% being 2,3-linked and 40% being 2,6-linked. Because of the restricted sialylation capacity of CHO cells, the recombinant EPO is sialylated entirely via the α2,3 linkages. Nevertheless, recombinant EPO produced from CHO cells is currently employed as a highly effective therapeutic agent in the treatment of a variety of renal dysfunctions. There is no evidence of any adverse physiological effect due to the structural difference in the terminal sialic acid.

Mouse cells express the enzyme α1,3 galactosyltransferase, which generates Galα1,3-Galβ1,4-GlcNAc residues not present in humans and is an epitope found to be highly immunogenic. The epitope is found in the glycan structures of glycoproteins from nonprimate animals including rodents, pigs, sheep and cows as well as New World monkeys. Although there is evidence for the presence of the gene of this enzyme in CHO and BHK cells, there appears to be no activity of the α1,3 galactosyltransferase enzyme. The potential for immunogenicity associated with this epitope limits the use of murine cells for therapeutic glycoprotein production.

Table 10.1. Characteristics of the glycosylation capacity of two widely used mammalian cell lines compared to normal human cells

CHO cells

- lack α2,6 sialytransferase activity. Sialyation by CHO cells only CHO cells only produces α2,3 linkages to terminal N-glycan structures
- unable to sulfate GalNAc residues. This is important in certain hormones
- lacks α1,3 fucosyltransferase. This prevents the formation of peripheral fucose residues (Fuc-α1,3 GlcNAc)
- lacks β1,4 N-acetylglucosaminyltransferase III. This prevents the formation of GlcNAC attachment to the core mannose to form a bisected glycan

Murine C127 cells

- lacks α2,3 sialyltransferase activity. Sialylation by C127 cells only produces α2,6 linkages to terminal N-glycan structures
- expresses α1,3 galactosyltransferase activity. This results in Gal-α1,3-Gal which is antigenic in humans

Genetic mutations can also affect glycan synthesis. A series of CHO clones has been isolated possessing a variety of mutations affecting N- and O-glycosylation. The mutants are characterized by the expression of aberrant lectins on the cell surface and are classified as a series of numbered LEC mutants. These mutations usually diminish the glycosylation capability. For example, Lecl CHO mutant expresses no detectable GnT 1 and accumulates glycoproteins with $Man_5CHcNAc_2$ structures in the cell. A mutant affected in the same gene, LeclA was isolated from a sub-population of this mutant. This mutant produces a GnT 1 biochemically different from the one produced by the parental cell line, with new kinetic properties. Multiple enzymic defects may be an advantage in the production of glycoproteins with minimal carbohydrate heterogeneity: A mutation may also result in a mutant with a gain of function such as the CHO Lec11, which expresses α1,3 fucosyltransferase. CHO cells have a limited capacity for synthesis of elongated O-glycans because of the lack of core 2 GlcNAcT activity, although this enzyme may be induced by butyrate treatment.

Culture Environment

The control of the culture environment is important to maximize cell growth in order to attain a high cell density, which is a prerequisite for producing cell products whether they be viruses, antibodies or recombinant proteins. However, it is also important to realize that the specific conditions of the culture can affect product glycosylation independently of the characteristics of the cell line. During the course of a batch culture nutrient consumption and product accumulation change the cellular environment, usually in a way that can gradually decrease the extent of protein glycosylation over time. Such changes are unacceptable in a cell culture bioprocess used for largescale production of a protein that may be a therapeutic agent. It can lead to variable glycoform heterogeneity and significant batch-to-batch variation in the production processes. In order to maintain product consistency it is essential to understand the parameters of cell culture that can cause variations in glycosylation. As a more far-reaching objective it may be reasonable to control culture conditions in favor of reducing glycoform heterogeneity or producing a specific glycoform. For example, the maximization of product sialylation of therapeutic proteins could lead to higher specific biological activities *in vivo*.

Two lines of evidence suggest that the extracellular environment may affect glycosylation: (a) *in vivo* changes in glycosylation are associated with the physiological state. e.g. pregnancy and disease (e.g. diabetes); (b) *in vitro* cell culture studies show direct effects of the extracellular environment on protein glycosylation. In some cases these have been reported from changes in the mode of culture. For example, the glycosylation of antibodies was found to be more consistent by *in vitro* culture than from ascites fluid or from the adaptation of cells from serum to serum-free medium. In other reports, the specific culture parameters affecting an alteration in glycosylation have been analyzed.

The choice of culture method, pH, nutrient concentration, dissolved oxygen, etc., are some of the parameters that may affect the glycan structures of glycoproteins. Among the potential mechanisms to explain such effects are:

1. Depletion of the cellular energy state;
2. Disruption of the local ER and Golgi environment;

3. Modulation of glycosidase and glycosyltransferase activities;
4. Modulation of the synthesis of nucleotides, nucleotide sugars and lipid precursors.

The awareness of such effects on the glycosylation of proteins makes the use of more defined culture media and conditions a very important issue for the development of a pharmaceutical product with defined glycan structures and batch consistency.

Mode of Culture

CHO and BHK cells can be grown as anchorage-dependent cells in T-flasks or microcarriers. Alternatively, they can be adapted to suspension culture. This adaptation process leads to characteristic changes in the expression of cell surface proteins such as the integrins which affect viral susceptibility. Not surprisingly, the glycosylation process may also be affected by such changes. Watson et al. (1994) reported that the sialylation of N-glycans of a secreted protein from CHO was reduced in microcarrier culture compared to suspension. Gawlitzek et al. (1995) also showed an effect on glycan antennarity of a BHK-produced protein related to the mode of culture.

The presence or absence of serum in the culture medium also has a significant effect on glycosylation. This is not surprising given the variable concentrations of hormones and growth factors in serum and even in different formulations of serum-free media. Cells grown in SFM (serum-free medium) secrete a higher proportion of N-glycosylated and O-glycosylated protein with enhanced terminal sialylation and proximal fucosylation. This result was attributed to the presence of high activities of sialidase and fucosidase in serum.

Protein Productivity

Differences in growth rate, specific productivity and cell density among the bioreactors may cause variations in the pattern of N-linked glycan structures, especially in the degree of sialylation. This has been shown in experiments using different bioreactors established under similar conditions. It has also been shown that the rate of protein expression may affect glycosylation. In one set of experiments the slower transit time of the glycoproteins through the Golgi at lower temperatures was shown to increase the number of polylactosamine residues in the glycans.

Glucose

Low glucose concentrations may produce two distinct abnormalities in the synthesis of glycoproteins: attachment of aberrant precursors to the protein and absence of glycosylation at sites that are normally glycosylated. Both abnormalities are due to a shortage of glucose-derived glycan precursors. Glucose starvation may result in an intracellular energy-depleted state or a shortage of glucose-derived glycan precursors. Reduced site occupancy of 1g light chains was observed in mouse myeloma cells grown at a glucose concentration below 0.5 mM. Abnormal glycosylation of viral proteins is also observed at low glucose. In a chemostat culture of CHO cells Hayter et al. (1993) showed an increase in nonglycosylated gamma-interferon under glucose-limiting conditions. Pulsed additions of glucose restored normal glycosylation rapidly.

Ammonia

Glutamine is normally added to culture medium at a concentration of 2-10 mM. The glutamine provides an energy source for cells as well as being an essential precursor for nucleotide synthesis. However, glutamine is a source of ammonia accumulation in culture medium which arises from either thermal decomposition of the glutamine or from metabolic deamination or deamidation. The accumulated ammonia is inhibitory to cell growth and also has a specific effect on protein glycosylation. Ammonia is likely to cause a decrease in sialylation and a change in the number of antennae of the glycan.

There are two major changes to the intracellular environment caused by ammonia that could explain its effect on glycosylation.

1. An increase in pH of the Golgi. This could decrease the activity of selected enzymes.
2. An increase in the UDP-GNAc pool. This could compete with the transport of CMP-NeuAc into the Golgi that is necessary for sialylation.

pH

Under adverse external pH conditions the internal pH of the Golgi is likely to change, resulting in a reduction of the activities of key glycosylating enzymes. The pH of the medium has been shown to affect the distribution of glycoforms of IgG secreted by a murine hybridoma. It has also been shown that the maximum glycan occupancy of a protein occurs between pH values of 6.9 and 8.2.

Dissolved Oxygen Concentration

Oxygen plays a dominant role in the metabolism and viability of cells; it is a limiting nutrient in animal cell culture because of its low solubility in the Medium. It has also been shown to affect the pattern of glycans on a protein expressed by cells in culture. Monoclonal antibodies have a biantennary glycan structure at a site on the heavy chains that may show two (G2), one (G1) or no (G0) terminal galactose groups. A decrease in galactosylation may occur at reduced dissolved oxygen (DO) concentrations (10% DO), when mainly agalactosyl or monogalactosylated glycans occur while at higher oxygen concentration (50-100% DO) there is a higher proportion of digalactosylated glycans.

The mechanism for the effect of DO on galactosylation is unclear. One explanation is that reduced DO causes a decline in the availability of the UDP-Gal. This might arise due to a sensitivity to reduced oxidative phosphorylation in the production of UDP-Gal or as a result of reduced UDP-Gal transport from the cytosol to the Golgi. A second explanation is based on evidence that the timing and rate of formation of the inter-heavy chain disulfide bonds in the hinge region of IgG determine the level of Fc galactosylation. Thus the addition of galactose may be impeded by the early formation of the inter-heavy chain disulfide bond. Low DO in the culture may cause a perturbation in the oxidating environment of the ER and/or the Golgi complex and the disturbance may result in a change in the pathway of inter-chain disulfide bond formation.

An effect of DO has also been observed in CHO cultures, with enhanced sialylation of a recombinant protein at high oxygen levels.

Growth Factors/Cytokines/Hormones

There are many reports of hormones involved in the regulation of protein glycosylation *in vivo*. Up- and down-regulation of specific glycosyltransferases has been observed frequently in conjunction with hormonal induction of cell differentiation. Presumably, transcriptional control of glycosylation enzyme concentration is responsible for many of the effects on glycan processing. An example of glycosylation control *in vivo* is the cascade of events that occurs following the stimulation of the synthesis of thyrotropin by the tripeptide, thyrotropin-releasing hormone (TRH). This in turn promotes the synthesis and sialylation of thyroglobulin by thyroid cells. The glycosylation of transferrin is regulated by prolactin in rabbit mammary glands.

In cell culture dexamethasone can affect glycan structures in rat hepatocytes. Retinol and retinoic acid may play a role *in vivo* in epithelial cell differentiation and can be shown in culture to cause significant changes to protein glycosylation. This includes a shift from high mannose to complex glycans in chondrocytes and the extension of complex structures in mouse melanoma cells.

Exogenous IL-6 induces changes in the activities of intracellular GnTs including a reduction in the activity of GnT III and an increase in GnT IV and GnT V of a myeloma cell line that led to alterations in the glycan structure of the surface and secreted glycoproteins.

Extracellular Degradation of Glycans

Mammalian cells possess glycosidases that may be released extracellularly into the culture by cell secretion or upon cell lysis. These include fucosidases, β galactosidases, β hexosaminidases and sialidases that may accumulate in the extracellular medium. CHO cells possess a significant and stable sialidase activity that can accumulate in the culture medium and retains considerable activity at pH 7. The action of these enzymes on secreted glycoproteins that have a variable residence time in the culture may result in significant heterogeneity of glycoforms.

The extent of glycan degradation depends on many factors, including the level of extracellular activity, pH, temperature and time of the glycoprotein exposure to the enzyme. Bioprocesses that result in maintenance of high cell densities for long periods such as fed-batch or perfusion mode cultures may be particularly vulnerable to this type of glycan degradation. Early extraction of the product from the medium reduces the residence time of the glycoprotein in culture and may reduce glycoform heterogeneity.

Genetic Engineering of Mammalian Cells to Modify Glycosylation

Mammalian cell lines used for the production of glycoproteins may lack the enzymic profile to synthesize recombinant proteins that are glycosylated as authentic human proteins. This may be due to the lack of processing enzymes, the presence of alternative processing enzymes or through the expression of glycosidase activities in the mammalian host cells.

Metabolic engineering provides a promising tool to modify the characteristics of the host mammalian cells by enhancing cell productivity, protein quality and bioactivity and by modifying the glycosylation pathway to obtain a final product with advantageous properties.

Engineering Host Cells for New Glycosylation Properties

The two commonly used hamster cell lines, BHK-21 and CHO cells do not express α2,6 sialyltransferase, α1, 3 fucosyltransferase or β-1,4-N-acetylglucosaminyltransferase III activities. As these enzymes are found in normal human cells, the products of the hamster cell lines may not possess some of the glycan structures found typically in human serum proteins. Transfection of the cells with the gene of the lacking glycosyltransferase may correct such deficiencies.

α2,6 Sialyltransferase (α2,6ST)

Two different sialic acid linkages (α2,3 and α2,6) to the terminal Gal are found in N-linked glycans isolated from human glycoproteins. The human enzymes responsible for these substitutions are α2,3ST and α2,6ST. Both enzymes compete for the same substrate and a mixture of both linkages is often found in native human glycoproteins. However, CHO and BHK-21 cells only produce α2,3 sialylated glycan structures.

The sialyltransferase (α2,6ST) gene has been introduced into BHK cells that express recombinant proteins. The modified cells produced glycoproteins with an increased level of sialylation which included a mixture of 2,3/6 sialylated glycan structures.

Similarly, CHO cells have been co-transfected with genes for IFN-γ and α2, 6ST. The modified cells produced IFN-γ, 68% of which was sialylated with a α2,6 linkage. The overall extent of sialylation was doubled compared to the product of the cells without the α2,6ST gene. The addition of sodium butyrate enhanced the α2,6ST activity and increased the extent of sialylation and the proportion of α2,6 linked SA to 82%. However, sodium butyrate had no effect on the sialylation of the product of the cells without the α2,6ST gene insert.

α1,3 Fucosyltransferase (α 1,3FT)

This enzyme is required for the addition of a peripheral αl,3 fucose linkage to GlcNAc as found in certain human proteins. The co-expression of β-TP from recombinant BHK-21 cells with human α1;3FT successfully produced a glycoprotein, 50% of which had an α1,3-linked Fuc. However, a significant decrease in the degree of sialylation of Nglycans was observed. It, is suggested that this could be due to competition of αl,3FT with the endogenous α2,3ST for the same substrate. The α2,3 sialyltransferase is unable to sialylate fucosylated structures.

β1,4 N-acetyl glucosaminyltransferase (GnT III)

The bisecting GlcNAc residue plays an important role in the branching and elongation of glycan structures by restricting the action of other enzymes. GnT III is not expressed at significant levels in normal CHO cells so the synthesis of bisected glycans does not occur in these cells unless they are transferred with the GnT III gene.

A CHO cell line capable of producing bisected glycans on the glycan structure of IFN-β has been created. This structure has not been detected in native human IFN-β and the biological significance of this is unknown.

Immunoglobulin glycosylation is essential for complement fixation and antibody-dependent mediated cytotoxicity (ADCC). This is a lytic attack on antibody-targeted cells and is initiated after the binding of a lymphocyte receptor to the constant region (Fc) of the antibodies. This effector function may be essential for the therapeutic application of certain antibodies. Although human serum IgG contains low levels of bisecting GlcNAc, therapeutic antibodies containing bisected glycans may have an enhanced ADCC.

One example of this is a chimeric IgG1 (chCE7) anti-neuroblastoma engineered in CHO cells which showed enhanced ADCC as a result of the presence of bisected glycoforms in the Fc region. The antibody chCE7 was constructed by transfecting the CHO parental cell line with the GnT III gene under tetracycline regulation. Over-expression of GnT III led to a modified IgG containing a bisected glycoform. The maximal activity of ADCC correlated with a high level of Fc-associated bisected complex glycans. Thus, enhanced ADCC activity of chCE7 together with the capacity of this antibody to recognize neuroblastoma cells, makes it a suitable candidate molecule for the treatment of these tumours.

The effect of GnT III expression levels on glycan structures in CHO cells was predicted by a mathematical model based upon enzyme kinetic constants and mass balances associated with the production of 33 different N-glycan structures. GnT III has the potential to act upon at least seven independent glycan structures that result in either bisected complex and bisected hybrid glycans. The complex form is the required structure to maximize biological activity *in vivo*. Analysis of the competitive enzymic reactions in the central reaction network of the Golgi can be used to predict the activity of GnT III required to maximize synthesis of the bisected complex glycoforms.

Metabolic regulation of O-glycosylation

The predominant O-glycan structures formed in CHO cells are the core I type. However it has been shown by the simultaneous upregulation and down-regulation of key enzymes that this pathway can be altered. CHO cells that had been genetically engineered to express α1,3 fucosyltransferase (α1,3FT) were selected for the co-expression of a CMP-sialic acid: Gal β1,3GalNAc α2,3-sialyltransferase

(ST3Gal I) gene fragment set in the antisense orientation and the human-UDP-GlcNAc:Galβ1,3GalNAc-R β1,6-N-acetylglucosaminyltransferase (C2GnT). This coexpression resulted in an increase in the activity of the C2GnT enzyme and a decrease in the activity of the ST3Gal 1 enzyme. The effect of this coordinated change was to divert the O-glycosylation pathway from the formation of core I glycans to core II glycans. The significance of this is the formation of proteins containing sialyl-Lewis X glycan structures that mediate interaction with selectins and cell-cell adhesion.

Antisense RNA and Gene Targeting

An alternative for metabolic engineering of producer focuses on the direct manipulation of the expression of gene characteristics of proteins by the use of antisense RNA. This approach is studied in the form of T.W. an unwanted enzyme activity or to enhance the expression of proteins to improve the product quality.

One obvious target for this strategy in CHO cells is the soluble sialidase gene. A reduction of sialidase expression would improve the stability of secreted proteins in culture supernatant. The de-sialylation of therapeutic proteins reduces bioactivity because the resulting proteins with exposed terminal Gal residues are removed from the bloodstream by hepatocyte asialoglycoprotein receptors. Thus, the conservation of sialic acid in the glycan chains of glycoproteins is critical and must be maintained on the proteins during the production and purification processes.

Antisense expression of sialidase resulted in a 60% reduction of sialidase activity in the culture supernatant of CHO cells expressing DNAase. Although only an additional one mole of sialic acid per mole of protein was observed, this modest improvement in sialylation resulted in a dramatic effect on the serum clearance rate of the protein.

Antisense RNA targeting has proven to be a valuable means to revive silent genes or correct gene defects. More complete glycosylation of recombinant glycoproteins may be possible if the activities of endogenous glycosyltransferases are increased above normal levels.

Two strategies for the construction of antisense expressing cells are possible: (i) The creation of a universal host cell line which expresses the desired antisense RNA. Once the cell line is established, the product expression vector is introduced. The advantage of this approach is the availability of a universal cell line constitutively expressing antisense RNA. (ii) The introduction of the antisense expression vector into an existing recombinant host after growth and productivity parameters have been optimized. This reduces the possibility of modifying the antisense RNA.

CONCLUSION

The choice of the cell expression systems and the control of the production parameters at earlier stages of bioprocess development are key factors for ensuring the production of glycoproteins with consistent structures.

To optimize the glycoform distribution for a given glycoprotein produced by a given cell type it is important to understand the specific environmental factors affecting glycan structures and how to control these factors at the cellular level. Given the biological complexities of cell growth and metabolism, the cellular and environmental parameters that can be potentially altered are enormous. An increased awareness of the importance of pharmaceutical protein glycosylation has led to the increased importance of analysis of glycan structures.

Recent efforts in metabolic engineering are clearly justified in view of the demands for the production of proteins with a consistent glycoform profile and more cost-effective, high-productivity processes. Continuing efforts in metabolic engineering may lead to host cell lines capable of producing a restricted set of glycoforms for a specific glycoprotein, with enhanced bioactivity and reduced blood clearance rates.

PROTOCOL

Analysis of Glycans from In-gel Release of Protein Bands

Materials

0.5 M dithiothreitol (DTT)

100 mM iodoacetamide (IAA)

NuPage 10% gel with MOPS pH 7 as running buffer

Sodium bicarbonate buffer 20 mM, pH 7

PNGase F (Boehringer) 1 unit/ml

Dowex AG50 ×12 anion exchange resin

GlycoClean S cartridge (Glyko)

2-aminobenzamide (2-AB) labeling reagent (Glyko)

Procedure

1. Reduce a protein sample (50-70 μg) with DTT for 10 min.
2. Alkylate sample with iodoacetamide for 30 min.

3. Apply protein samples (5-10 μg) to individual gel lanes for separation by SDS-PAGE with precast gels and MOPS running buffer.
4. Stain gel with Coomassie Blue to reveal distinct protein bands.
5. Remove the protein bands from three identical lanes from the stained gel by scalpel. Cut into small pieces (1 mm^2) and transfer to an Eppendorf tube.
6. Freeze the tube for at least 2 h to break up gel prior to enzymic digestion.
7. Wash alternately with sodium bicarbonate buffer 20 mM, pH 7 and acetonitrile.
8. Treat the dried gel pieces with PNGase F enzyme (3 μl) and sodium bicarbonate buffer (27 μl) for 12-16 h at 37°C to remove the attached glycans.
9. Extract the glycans from the gel pieces by repeated (× 3) sonication with 200 μl water for 30 min.
10. De-salt the extract by passage through a 0.5 ml Dowex AG50 ×12 column and elution with 4 × 0.5 ml water. Dry the sample completely prior to labeling.
11. Label each sample by reductive amination with the fluorophore, 2-aminobenzamide (5 μl) at 5°C for 2 h.
12. Remove excess labeling reagent by acetonitrile elution through a GlycoClean S cartridge. The sample elutes subsequently with water. Alternatively, ascending paper chromatography (Whatman 3 MM filter paper) with acetonitrile can remove the reagent.
13. Each labeled glycan sample can be analyzed by normal phase HPLC.
14. Structural assignment of the peaks arising from this separation can be performed by subsequent treatment with multi-enzyme arrays of exoglycosidases.

11

ANTIBODIES IN CULTURE

Since the 1970s, monoclonal antibodies have become essential tools for most research laboratories involved in cell and molecular biology. They have also been used to develop routine diagnostic medical tests and more recently as therapeutic agents.

The purpose of this chapter is to describe the background to the methods used to produce hybridomas that secrete monoclonal antibodies (Mabs) and to show how this can be done in the laboratory. The techniques described can be performed in any cell culture laboratory with access to animal facilities. Alternatively, a wide range of antibody-secreting hybridomas is available from culture collections.

ANTIBODY PRODUCTION *IN VIVO*

Antibodies are found in a specific protein fraction of blood called the gammaglobulin or the immunoglobulin fraction. They are synthesized by a subset of white blood cells—the B-lymphocytes. The basic structural arrangement of two heavy associated with two light chains is similar for all the isotypes. However, each isotype is distinguished by different heavy chain structures, which are of varying length, number of domains and glycan structures.

Each B-lymphocyte is capable of producing one type of antibody in response to a particular antigen which interacts with a cell surface receptor. Stimulation by an antigen causes growth and an expansion of the cell population capable of producing the corresponding antibody. The variety of antibodies present in any animal reflects the population of B-lymphocytes which have been stimutlated by previous exposure to a range of antigens. A particular antibody will be produced in an animal following the injection of the corresponding antigen. For example, if human insulin is injected into a mouse, after a few days the blood will contain significant quantities of mouse antibody capable of binding to human insulin.

The immunoglobulin fraction of the mouse blood can be extracted and will contain the required anti-insulin. However, this fraction will also contain numerous other antibodies and it would be very difficult to isolate the particular antibody that may be required. Because of the multiplicity of immunoglobulin types in the fraction the term 'polyclonal antibody' is used. This polyclonal antibody may even include different antibodies against insulin, i.e. antibodies reactive to different regions (epitopes) of the insulin molecule. Antibodies are found in all body fluids including blood, milk and mucous secretions and serve an essential role in the immune system that protects animals from infection or the cytotoxic effects of foreign compounds. Antibodies will bind with high affinity to an invasive molecule. Normally the binding is to only part of a large molecule (the epitope) and so there may be many different antibodies for a particular compound. Antibodies are valuable because of their very specific recognition and affinity for one compound (the antigen).

Structure of Antibodies

This has an overall molecular mass of 150 kDa and is the major class of immunoglobulin found in blood serum. The molecular structure consists of two light and two heavy chains bound by disulfide bridges. The heavy chain of lgG has four domains VH-CHl-CH2-CH3 and the light chain has two domains VL-CL. The 'constant' region (C) of a particular immunoglobulin class varies only with the species of origin. For example, a human IgG would have a different constant region from a mouse IgG. The variable domains (V) account for the diversity of antibody structure. Digestion of the molecule with papain cleaves the heavy chain in the 'hinge' region and results in three fragments. Two Fab (antibody-binding fragments) each contain the N-terminal end of a heavy chain with disulfide-linked light chain. The other fragment is the Fc which consists of the C-terminal end of the two heavy chains.

There are two glycan structures present in the space between the two CH2 domains. In some immunoglobulins there are also glycans present in the variable region of the molecule. Each immunoglobulin molecule has a unique antigen-binding region which consists of hypervariable sequences of amino acids. This region enables the immunoglobulin to bind to one specific molecule (the antigen) with

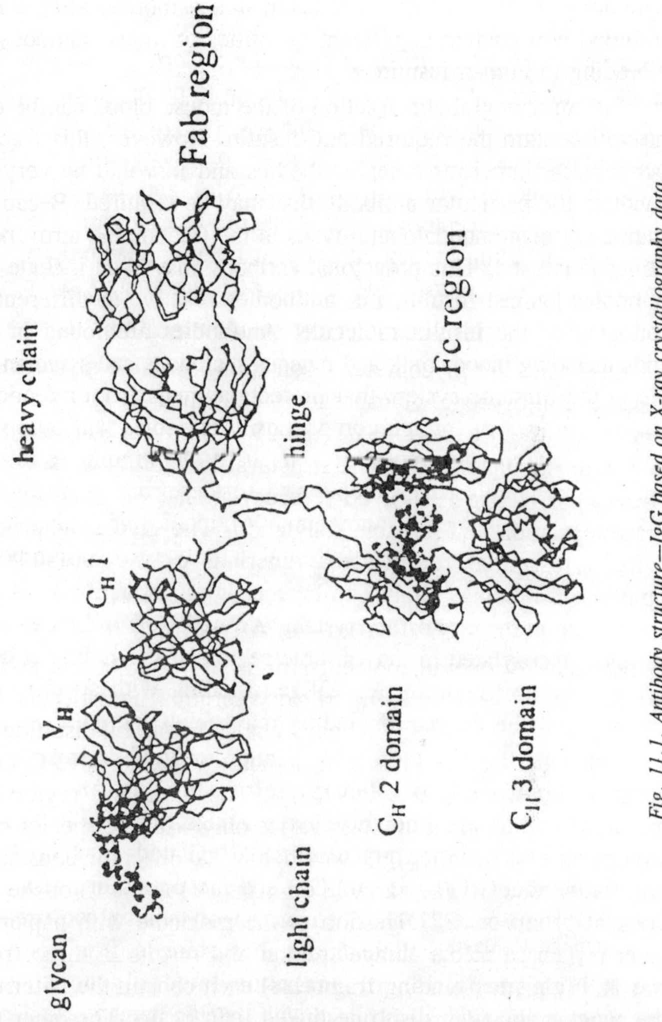

Fig. 11.1. Antibody structure—IgG based on X-ray crystallography data.

high affinity. The unique antigen-binding site which consists of hypervariable sequences of amino acids is the Fab region. These are formed from three hypervariable loops (complementarity determining regions, CDR) of the VH domain and three hypervariable loops of the VL domain. The variable sequence is produced by somatic recombination and by mutagenesis and accounts for the diversity of antibody molecules. This region enables the antibody to bind to one specific molecule (called the *antigen*) with high affinity.

The other important functional component of the molecule is the effector site which is found in the constant region. The effector functions can be mediated by the Fc region binding to complement (Clq) and to receptors of specific cells. Complement activation leads to the activation of leukocytes and phagocytosis. The Fc receptors are on certain cells of the immune system such as phagocytes and natural killer (NK) cells. Binding to receptors in these cells produces a variety of biological responses including antibodydependent cellular cytotoxicity (ADCC), phagocytosis, endocytosis and release of inflammatory agents.

Glycosylation of Antibodies

Antibodies are glycoproteins containing variable glycan structures. A single conserved N-glycan site is contained in IgG on each CH2 domain of the Fc region at Asn-297. The carbohydrate is of a biantennary complex type. The structural variability is associated with a bisected GlcNAc (+/–), core fucosylation (+/–), non-, mono- or digalactosylation and possible sialylation. The glycosylation of the Fc region is essential for effector functions of the antibody such as complement binding, binding to Fc receptors, induction of ADCC and the half-life in the circulatory system. Around 20% of human antibodies are also glycosylated in the variable region of the Lab fragment. This glycan may be important for antigen binding with specific examples showing that the degree of binding may either increase or decrease.

Although the level of glycosylation of IgG is small (2-3% by weight), compared to other proteins, the glycan structures on immunoglobulins are known to have a significant effect on immune responses. The common glycoforms of IgG with 0, 1 or 2 galactose terminal residues (G0, G1 and G2) and the possibility of a sialic acid terminal group on G2. The distribution of these glycoforms changes under certain pathological conditions. For example, it is well established that a high proportion of agalactosylated glycan structures in immunoglobulin (G0) is associated with specific human disorders, notably rheumatoid arthritis. Here the predominant form of the glycan attached to Asn-297 is a biantennary complex structure that terminates in N-acetylglucosamine residues and lacks the usual galactose terminus. This altered glycan structure results in a change in the interaction of the immunoglobulin with specific monocyte receptors. Also, there are changes to the structure of the immunoglobulin caused by the altered glycan that may explain changes in immune response related to the disease condition.

Monoclonal Antibodies

In the 1970s methods were developed to produce antibodies from cells in culture. In 1975 Choler and Milstein described a technique to produce cells which were hybrids of B-lymphocytes and myelomas and which were capable of continuous synthesis of preselected antibodies. Myeloma cells are transformed lymphocytes capable of growing indefinitely. Since the 1970s, their method has found wide application and has resulted in the large-scale production of kilogram quantities of some monoclonal antibodies. The term 'monoclonal' indicates that the antibody is of a single type i.e. will bind to just one antigen. Kohler and Milstein obtained the Nobel Prize in 1984 for their work on the development of Mab-secreting hybridomas.

Hybridomas

Hybridomas are hybrid cells capable of the continuous production of monoclonal antibodies. They combine the key properties of the two parental cells: a myeloma with an infinite lifespan and a B-lymphocyte capable of synthesizing a single antibody. Hybridomas can be grown in suspension in large bioreactors for the production of kilogram quantities of monoclonal antibodies. The antibodies have a range of applications because of their high specificity in recognizing selected proteins. This enables them to be used for diagnosis and testing in applications such as blood typing, the detection of virus, pregnancy testing or for the detection of contaminants in food. The application of monoclonal antibodies as human therapeutic agents in the treatment of disease has been suggested for a number of years. However, there have been difficulties in the production of antibodies that are not immunogenic to humans. In the late 1990s a range of human or 'humanized' antibodies have been produced specifically for the treatment of cancer. The number of such therapeutic monooclonal antibodies is likely to increase in the future as a result of the numerous clinical trials that are now taking place.

The original work of Kohler and Milstein described, the creation of a mouse-mouse hybridoma that secreted antibody with affinity to sheep red blood cells as antigens. The antibody could be easily detected by a hemolytic plaque assay that showed the capability of the antibody to bind to and lyse sheep erythrocytes. Since the 1970s, these methods have found wide application and have resulted in the large-scale production of kilogram quantities of some monoclonal antibodies. The term 'monoclonal' indicates that the antibody is of a single type. This will bind to just one antigen.

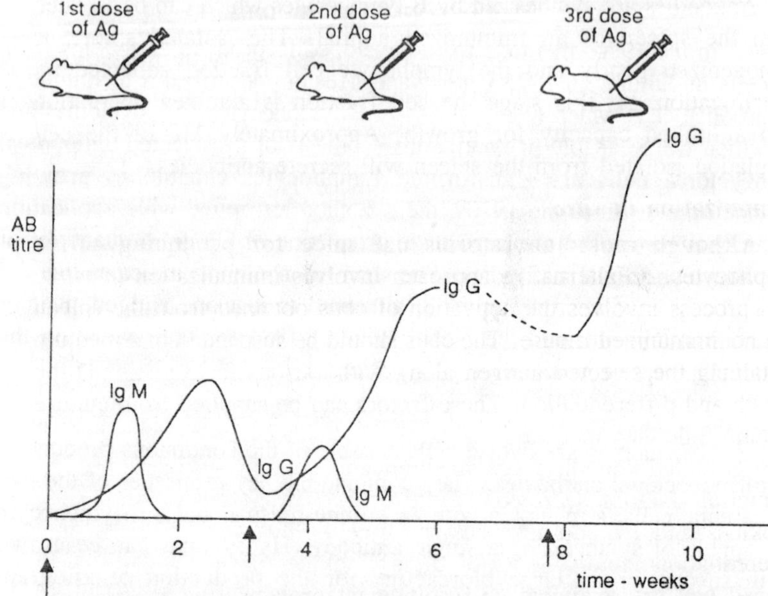

Fig. 11.2. Boosting the antibody response by multiple immunizations with a specific antigen (Ag).

The method developed by Kohler and Milstein involves four stages which result in the production of a hybrid lymphocyte with an infinite growth capacity and capable of continuous synthesis of a single antibody.

(i) immunization;

(ii) cell fusion;

(iii) genetic selection;

(iv) cell selection.

Immunization *in vivo*

The first stage of the production of a hybridoma is to obtain lymphocytes from an animal that is enriched with specific antibody-secreting cells. Immunization involves the injection of a chosen antigen into an animal—mice and rats have been most commonly used. The time required to produce an immune response resulting in antibody synthesis will depend upon the antigen, but a period of up to 3-4 weeks ensures maximum response. Large molecules tend to produce a strong response over a short period of time. Small molecules are often conjugated to carrier proteins such as albumin and multiple injections spaced over several days may be necessary to enhance the immune response.

Antibodies are synthesized by B-lymphocytes which can be isolated from the spleen of an immunized animal. The isolated spleen is homogenized gently and the lymphocyte cell fraction collected by centrifugation. At this stage the cell fraction is a mixed population with a limited capacity for growth. Approximately 1% of the cell population isolated from the spleen will secrete antibodies.

Immunization *in vitro*

Although most laboratories use mice for producing active lymphocytes, an alternative approach involves immunization *in vitro*. This process involves the activation of cells obtained from the spleen of a nonimmunized mouse. The cells should be suspended in a medium containing the selected antigen along with various factors stimulating growth and differentiation. These factors can be supplied from culture medium following incubation with mixed lymphocytes (or thymocytes). This is called 'conditioned' medium and contains various growth-promoting factors secreted by the cells. These factors are called cytokines and include interleukins, B-cell growth factor and B-cell differentiation factor. Some of these cytokines have now been well defined and are available as recombinant proteins from commercial suppliers. The effectiveness of immunization *in vitro* is dependent upon the optimal combination of these factors during cell activation.

An advantage of immunization *in vitro* is that the activation of B-lymphocytes takes 3-4 days rather than a few weeks as is the case with immunization *in vivo*. Furthermore, weak antigens at low concentrations can be used. A disadvantage is that certain immunoglobulin isotypes tend to be produced prefetentially (usually IgM) although refinement of the techniques during cell activation can stimulate the production of other isotypes.

Cell Hybridization—Process of Fusion

In 1965 Harris and Watkins reported that inactivated Sendai virus caused the hybridization of a mixed population of human HeLa cells and mouse Ehrlich ascites tumour cells. The result of the fusion was a mixed population of hybrid cells (called *heterokaryons*) that were genetically unstable. The sequence of events during fusion showing the cytoplasmic fusion of two dissimilar cells followed by the hybridization of the nuclei of the two cells. After a period of growth the heterokaryons tend to lose some of their genetic material and become stable hybrids retaining some of the phenotypic characteristics of each parental cell. The method turned out to be an extremely valuable tool for biological research, particularly in developing an understanding of

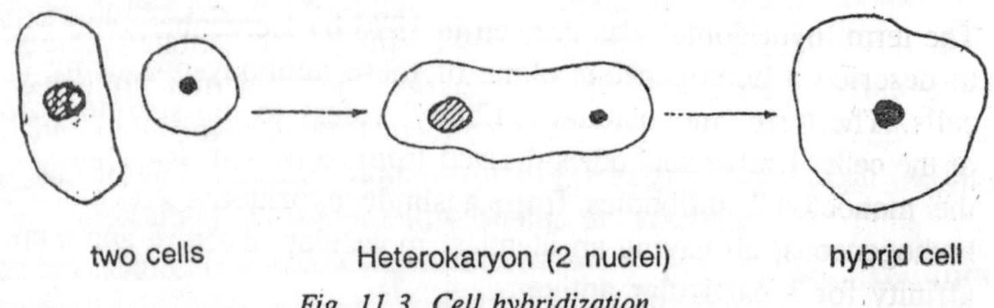

Fig. 11.3. Cell hybridization.

gene regulation. In 1969 Harris showed that when normal cells were fused with malignant cells the malignant phenotype was not always retained. This was the first direct evidence for the existence of human suppressor genes, derived from the normal cells and that could result in suppression of the tumourigenic characteristics. These genes, whose product include the retinoblastoma protein and p53, are now well characterized in terms of their role in malignancies. Cell fusion has also been used extensively in human chromosome mapping. The heterokaryons resulting from the fusion of human and mouse cells are genetically unstable and tend to lose human chromosomes randomly. This eventually gives rise to a mixed population of stable hybrids from which individual cell clones can he isolated.

Many of these clones may contain single human chromosomes. It is the association of a particular chromosome in an isolated cell clone with a selected measurable phenotypic characteristic such as an enzyme activity that allows the gene of that enzyme to be assigned to a specific human chromosome. With the use of this technique, many human genes have been assigned to particular chromosomes (known as 'chromosome mapping'). It was this same technique of cell fusion that Kohler and Milstein used in their work reported in 1975 that allowed the creation of stable hybrid cells from the hybridization of antibody-secreting B-lymphocytes with transformed myelomas.

The resulting cells retained two important phenotypic characteristics from the parents—the ability for infinite growth (from the myeloma) and the ability to synthesize antibody (from the lymphocyte). The original objective of this work was to study somatic mutation as a mechanism for antibody diversity. This is the ability of B-lymphocytes to go through a maturation process following initial contact with an antigen to produce antibodies of increasing affinity. However, the application of the cell fusion technique to produce antibody-producing cells with an infinite growth capacity had a major impact on the ability to produce large quantities of antibodies that could be used for a variety of functions both in biological research and also as medically important products.

The term 'hybridoma' was derived in 1976 by Herzenberg and Milstein to describe a homogeneous clone of these antibody-producing hybrid cells. The term 'monoclonal antibody' refers to the secreted product of the cells. Unlike antibodies derived from blood samples ('polyclonal'), the monoclonal antibodies from a single hybridoma are molecularly homogeneous, all having an identical molecular structure and a specific affinity for a particular antigen.

Cell Fusion to Immortalize Lymphocytes

Although immunization can result in lymphocytes capable of producing the required antibody, the cells will only grow for a limited period of time. The purpose of lymphocyte hybridization is to combine the desired property of antibody synthesis of the B-lymphocyte population with the infinite growth capacity of a myeloma. Therefore, the selected lymphocytes are fused with a population of myeloma cells. Suitable myeloma fusion partners are selected for two other important characteristics:

1. Nonproduction of antibodies. This is desirable so that the resulting hybridoma does not synthesize more than one antibody;
2. Possession of a genetic marker, such as the lack of an enzyme, to allow cell selection. For example, myelomas deficient in HGPRT (hypoxanthine guanine phosphoribosyl transferase) are commonly used. This allows selection of hybridomas in HAT (hypoxanthine, aminopterin, thymidine) medium.

Table 11.1. Commonly used rodent cell lines (myelomas) as fusion partners

Species	*Cell line*	*Immunoglobulin expression*
Mouse	X653	No
	N50	No
	Sp2/0	No
	NSI	Yes
Rat	YB2/0	No
	Y3-Ag	Yes

To allow fusion the activated B-lymphocytes are mixed with a suitable fusion partner—the myeloma.

Methods of Cell Fusion

Cells can be induced to fuse if two cell populations are brought close together at a high cell concentration (10^6 to 10^7 cells per well of a 96 multiwell plate) in the presence of viruses or by chemical agents (called '*fusogens*'). The process involves the destabilization of adjacent cell membranes which eventually fuse to form a hybrid cell. Initially, two distinct nuclei are present in the fused cell (a *heterokaryon*). Eventually the nuclei fuse to produce a stable hybrid cell.

Although UV-inactivated Sendai viruses were originally used as agents for cell fusion, the more widely used method is now fusion by the chemical agent polyethylene glycol (PEG). This is a polymer, available at a molecular weight range of 200-20,000 kDa. PEG at 4000-6000 kDa is most suitable for cell fusion. Cell fusion can occur in a solution of PEG (40-50% w/v) within 1-2 minutes. In this process, cell swelling accompanies fusion. This enables adjacent cells to approach very closely and also the plasma membrane becomes permeable to small ions. However, lysis of swollen cells may also occur.

Alternatively, electrofusion can be used. In this technique, two populations of cells are introduced into a small sterile chamber. An electric current is applied in high-voltage pulses for short time periods. During this period the membrane will become highly permeable. This is similar to the process of electroporation used to facilitate the entry of DNA into cells. This causes the cells to orientate along the line of the current and fuse. This process is highly efficient, producing a high percentage of viable hybrid cells. The most suitable voltage for electrofusion is one that causes approximately 50% death in the cell population. This would typically be around a voltage of 200 V for a cell pellet held in a small electroporation cuvette. From such a protocol there may be around 50 fused cells from an original total of 5×10^6 cells from each parental cell line.

Selectable Gene Markers for Cell Selection

The process of cell fusion will result in a heterogeneous population of cells that will contain unfused parental cells, lysed cells as well as the required hybrid cells. At this stage, cell selection is important so that the hybrid cells can be isolated from the mixture. For hybridomas there are two important stage of cell selection:

1. Isolation of hybrid cells from parental cells;
2. Selection of antibody-secreting cells within the hybrid cell population.

The basis of cell selection is to distinguish cell types through genetic differences. One required characteristic of the resulting hybridomas is the ability for effective growth in culture. So, initial cell selection can involve the incubation of the mixed population of cells in a suitable culture environment. This includes the addition of a suitable liquid growth medium to the cells and incubation at 37°C for a few days. This will allow the growth of all viable cells which will include the hybridomas and myelomas. Growth of these cells will dilute any nonviable and lysed cells front the mixture. Unfused lymphocytes have only a limited capacity for growth and will be eliminated eventually.

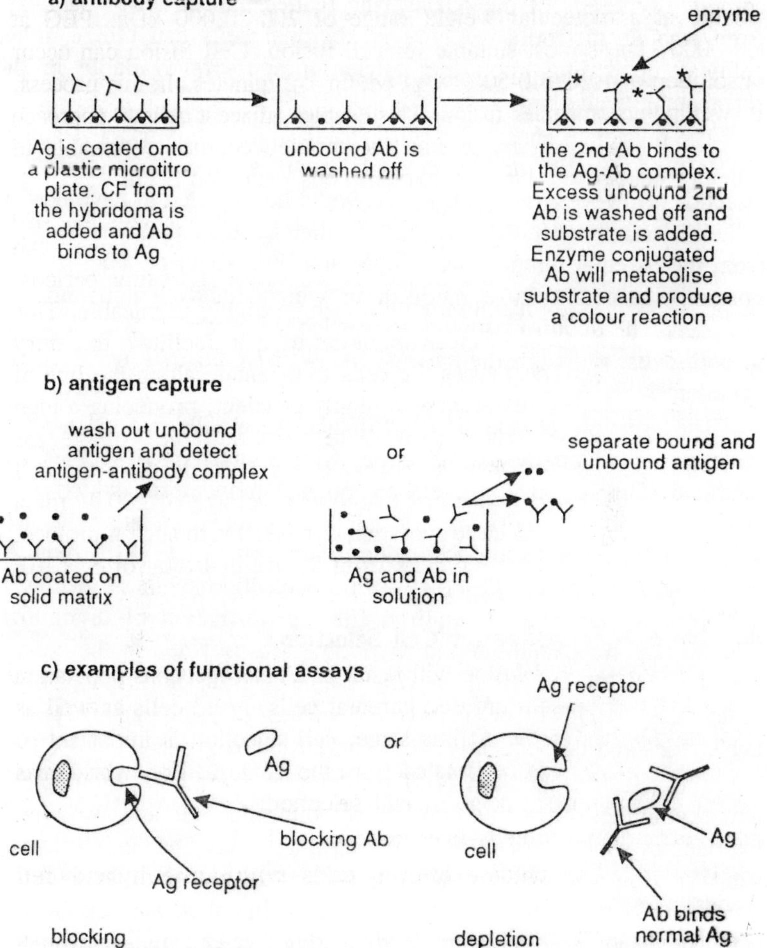

Fig. 11.4. Antibody screening assay. (a) Antibody capture; (b) Antigen capture; and (c) Functional assays.

The hybridomas are selected from the myelomas on the basis of a genetic marker that is normally applied to the myelomas. The most commonly used genetic marker is HGPRT which indicates a cell with a defective enzyme, hypoxanthine guanine phosphoribosyl transferase. This is a normal metabolic enzyme that is capable of catalyzing the addition of phosphoribosyl pyrophosphate on to hypoxanthine or guanine. This constitutes the salvage pathway for converting purines to nucleotides as part of nucleotide synthesis in all normal cells. However, there is an alternative pathway of nucleotide formation which involves the complete synthesis of the purine ring (called the '*de novo*' pathway) from smaller metabolic precursors. Normal cells can utilize either pathway depending upon the nutrient precursors available in the surrounding media.

The myelomas chosen for cell hybridization with lymphocytes have an HGPRT genetic marker through previous random mutation and selection. However, the hybridomas would be expected to have normal HGPRT activity ($HGPRT^+$) because they receive the normal gene from the parental lymphocyte. Either cell line would be able to grow normally in cell culture medium in which standard nutrients are provided. The *de novo* pathway for nucleotide synthesis would operate in both cells, although the salvage pathway would only occur in the hybridomas.

The principle of selection is to place these cells into a 'selective medium' in which the *de novo* pathway of nucleotide synthesis is inhibited. The key to this is the compound, aminopterin, which is an analog of folic acid and a specific inhibitor of dihydrofolate reductase, an essential enzyme for the formation of tetrahydrofolate (FH_4) required as a co-enzyme of the *de novo* purine nucleotide synthesis pathway. Tetrahydrofolate is also required for the formation of thymidine. However, if hypoxanthine and thymidine are provided in the culture media of $HGPRT^+$ cells they will be able to grow normally. On the other hand $HGPRT^-$ cells have no means of synthesizing purine nucleotides and consequently would be unable to grow.

The selective medium HAT contains hypoxanthine, aminopterin and thymidine. This medium allows the selection and growth of hybridomas which are $HGPRT^+$ and have a normal functioning salvage pathway. However, HAT is unable to support the growth of the $HGPRT^-$ myelomas because the *de novo* pathway is inhibited and the salvage pathway cannot function because of the defective enzyme. In the mixed population of cells resulting from fusion, the newly formed hybridomas will be $HGPRT^+$ by inheritance from normal lymphocytes whereas the unfused myelomas carry the mutant $HGPRT^-$ marker. Therefore incubation in HAT will allow survival of the $HGPRT^+$ hybridomas but not the $HGPRT^-$ myelomas. So, after a few days incubation in HAT the culture will contain only hybridomas.

Clonal Selection of Mab-secreting Hybridomas

After genetic selection with HAT the culture contains hybridomas but only some of these will secrete antibodies. Although the efficiencies of synthesis vary considerably, only about 10% of the population of hybridomas formed from cell fusion would be expected to secrete antibody. The next stage involves selection, of Mab-secreting hybridomas from the population which has survived HAT treatment. Cell clones can be isolated by the method of limiting dilution.

Cloning ensures that all cells selected for future cultures are genetically identical. This involves dispensing a cell suspension into a 96-well plate, so that each well contains an average of one cell. Hybridomas grow poorly at low densities but growth can be supported by a feeder layer of cells. This consists of a population of cells (e.g. thymocytes, macrophages or splenocytes) which have been treated to prevent any growth. The treatment may be by gamma irradiation or exposure to mitomycin C which halts DNA synthesis. Typically cells would he treated with 10 μg/ml mitomycin C for 2 h at 37°C. Under these conditions the cellular metabolism continues and secreted growth

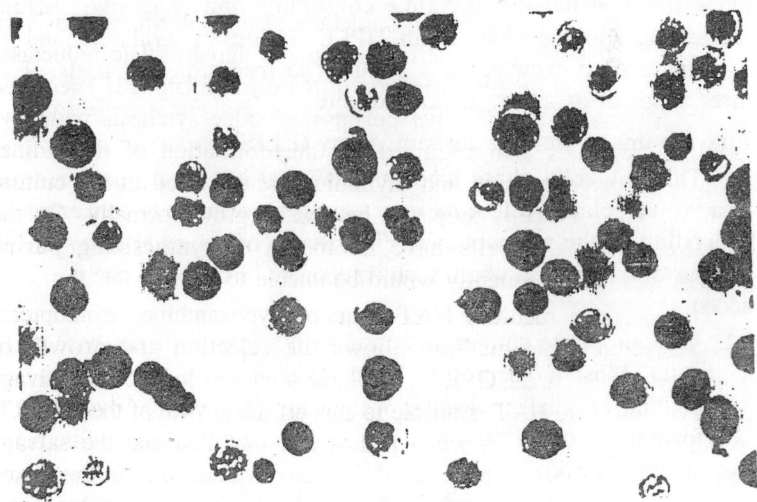

Fig 11.5. Hybridomas in suspension during exponential growth.

factors can promote the growth of viable hybridoma cells particularly if they are inoculated at low density.

Feeder layers can be purchased as frozen suspensions. After allowing 1-2 weeks for growth, the medium of each well should be tested for antibody content using a suitable assay. Wells containing a high antibody titer are then selected for further cell growth. At this stage the cells may be genetically unstable and a second round of cloning is recommended to ensure the isolation of a stable population of high-level antibody-secreting cells. This involves further testing of the culture medium for antibody content. Antibody-secreting hybridomas can be grown in culture in suspension to produce 1-2 × 10^6 cells/ml and a Mab concentration of 100-200 μg/ml. The cells can generally be adapted to serum-free medium while maintaining high productivity.

Assay of Monoclonal Antibodies

There are three assay procedures that are commonly used to detect monoclonal antibodies in solution. Each is suitable for the measurement of Mab concentration in culture media.

- ELISA—enzyme-linked immunosorbent assay.
- RIA—radioinnnunoassay.
- Affinity chromatography.

ELISA is the most commonly used assay for antibodies and is adapted to multi-well plates for analyzing multiple samples. RIA is more sensitive but is more time consuming and expensive. Affinity chromatography is ideal if an HPLC system is available and the hybridomas are growing in a serum-free medium. The basis of the three types of assay are described here.

Enzyme-linked Immunosorbent Assay (ELISA)

This is a solid phase binding assay that can easily be performed in a 96-well plate. ELISA measures antigen or antibody concentration, depending on the protocol used. The stages of a typical assay involve a series of additions in which each component binds to the one previously added.

1. An appropriate antigen is bound to the plastic of the base of the plate. Most large proteins bind spontaneously. Difficulties with binding of small molecular weight antigens can be overcome by forming a conjugate with a larger molecule such as bovine serum albumin (BSA).
2. The remaining attachment sites are blocked on the solid support by addition of a non-interfering protein such as BSA.

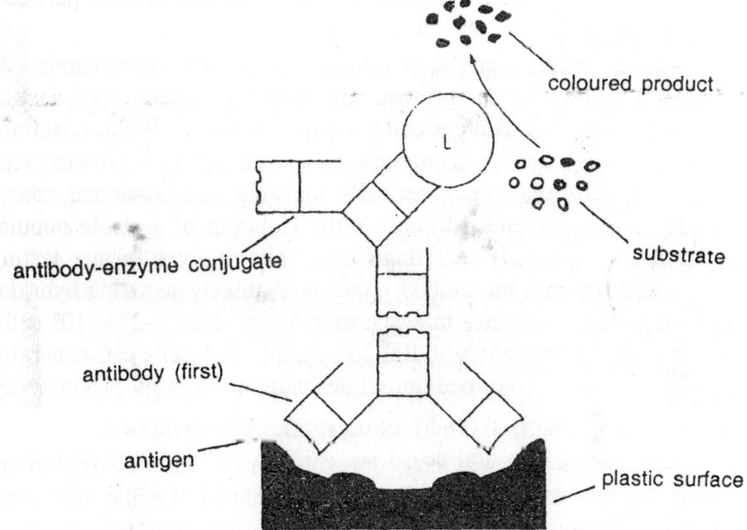

Fig. 11.6. Enzyme-linked immunosorbent assay (ELISA).

3. A solution of the Mab under test or a standard antibody solution is added. This will bind to the antigen held on the solid support.
4. An antibody-enzyme conjugate with specificity against the first antibody is added. The second antibody is species specific. This means that it binds to any immunoglobulin of the species from which the first antibody is derived, e.g. goat anti-mouse antibody. An enzyme such as alkaline phosphatase is conjugated to the second antibody which can then be detected by a colorimetric assay. Conjugated anti-Ig antibodies are available commercially.
5. Finally a suitable enzyme substrate is added. The substrate is one which can be changed to a coloured product by the enzyme bound to the conjugate. For example, p-nitrophenyl phosphate would be suitable for an alkaline phosphatase conjugate. The extent of colouration in each well of a plate can be measured by a multi-well reader.

Radioimmunoassay (RIA)

A solid phase binding assay similar to ELISA can be adapted to radioactive detection when a radioactively labeled antigen or antibody is used. The RIA is more sensitive and reliable than ELISA but is usually more time consuming and more expensive because of the cost of the radioactive label.

Affinity Binding

Certain bacterial cell wall proteins (called *Protein A* and *Protein G*) bind with high affinity to mammalian immunoglobulins. Protein A is derived from *Staphylococcus aureus* and has a strong affinity for antibodies. This allows antibodies to be isolated by chromatography columns which contain inert, beads conjugated to Protein A or G. If a sample (such as Mab-containing culture medium) is run through the column at neutral pH only antibodies will bind, allowing all other components to be washed out. Pure antibody can then be eluted from the column by a low pH buffer.

Suitable affinity columns of this type have been designed for use with high-performance liquid chromatography (HPLC) and this offers an extremely rapid method of analyzing or purifying antibodies. However, the method will detect any mammalian immunoglobulin which means that the immunoglobulin content of the serum used in the growth medium may interfere with analysis. HPLC affinity columns (such as ProAnaMabs from Hyclone) offer a rapid assay for measuring antibodies in serum-free culture medium and they could be used instead of ELISA. Affinity chromatography can also be used for large-scale antibody

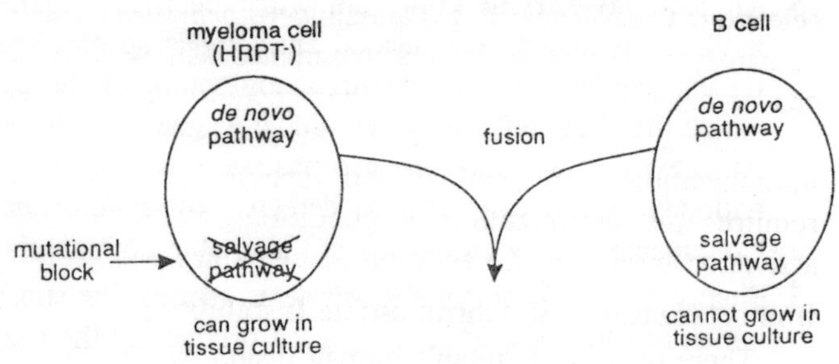

de novo pathway blocked by drug using HAT medium

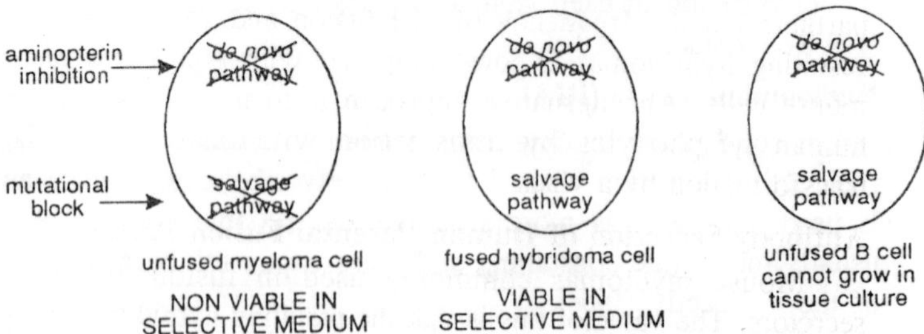

Fig. 11.7. Drug selection for viable hybridomas.

extraction, although the preparative Protein A or G columns are expensive.

HUMAN MONOCLONAL ANTIBODIES

Although murine-derived monoclonal antibodies are widely used as laboratory reagents, in affinity purification and clinical diagnostic tests, they have had limited success in human therapy. Immunoglobulins synthesized from mice and humans have different constant regions and so any antibody of mouse origin injected into a human could elicit an undesirable immune reaction. Firstly, although the antigen-binding site might be appropriate for the target, the antibody will not produce appropriate human effector responses such as those of complement and Fc receptor binding. Secondly, the human immune system will produce antibodies against the murine immunoglobulin. This is referred to as the human anti-murine antibody (HAMA) immune response. This presents an obstacle in developing therapeutic antibodies from a murine source. However, there are at least three major difficulties in producing human hybridoma cells capable of secreting human monoclonal antibodies.

Source of Antibody-secreting Lymphocytes

In generating murine hybridomas, the spleen of an immunized mouse is used as a source of the mixed lymphocyte population for cell selection. Clearly this is not possible with humans and the source of human lymphocytes is limited to samples of peripheral blood. These can be taken from patients who have acquired an immunity against a particular compound or disease. Alternatively, methods of *in vitro* immunization of human lymphocytes are possible. This approach requires the optimization of conditions for human B-lymphocyte activation by use of the appropriate cytokines and growth factors.

Immortalization and Chromosome Instability

There must be a suitable human fusion partner to immortalize the B-lymphocytes. Human myeloma cell lines are difficult to grow in culture. Human lymphoblastoid cell lines have been used as fusion partners but the frequency of cell fusion and genetic stability of the resulting hybridomas is low compared with equivalent fusions with mouse cells. An alternative approach is to immortalize the activated human lymphocytes by transfection with oncogenic DNA or by transformation by a virus.

Antibody Secretion of Human Parental Fusion Partners

Mouse myelomas commonly used in fusion are nonantibody secretors. The value of this is that the resulting hybridomas only secrete the antibody associated with the fused B-lymphocyte. Therefore, the culture product will be a single selected antibody type. However, most of the human myeloma or lymphoblastoid cells commonly used for hybridization are immunoglobulin secretors. This means that the resulting selected human hybridoma will secrete at least two antibodies, which are those associated with each of the parental cells.

Table 11.2. Human or human hybrid fusion partners

Cell type	*Cell line*	*Immunoglobulin Expression*
Human myeloma	SK007	Yes
Human lymphoblastoid	RH-L4	Yes but nonsecretor
Human lymphoblastoid	GM1500	Yes
Human lymphoblastoid	KR4	Yes
Human lymphoblastoid	LICR-LON	Yes
Human/human hybridoma	KR12	Yes
Human/mouse hybridoma	SHM-D33	Yes but nonsecretor

Recombinant Antibodies

A further possibility is the humanization of monoclonal antibodies originally produced from mice. This process involves antibody engineering, which relies on the techniques of recombinant DNA technology to rearrange some of the molecular domains of an immunoglobulin. In a chimeric antibody the mouse variable regions are linked to human constant regions. Thus in such a construct the antigen-binding site of the murine antibody is retained but the human constant region contributes the immunogenicity through the effector functions. A further step to humanizing the antibody is by replacing portions of the V-region that are not required for the antigenbinding site. The framework regions (FR residues) which were originally murine are replaced by human regions. Thus only the complementarity determining regions (CDRs) are retained as of murine origin.

Hybrid antibodies of this type have now been used as human therapeutic agents. The elimination of the murine constant regions reduces the previously experienced HAMA response. It is not always certain that complete humanization has an advantage over a chimeric antibody because humanization of the V-region may result in a loss in affinity to the antigen. Also, it is not clear that the problem of unwanted immunogenicity can be totally removed because repeated doses of even a fully humanized antibody may elicit an anti-idiotype response, that is directed against the antigen-binding site. However, these developments

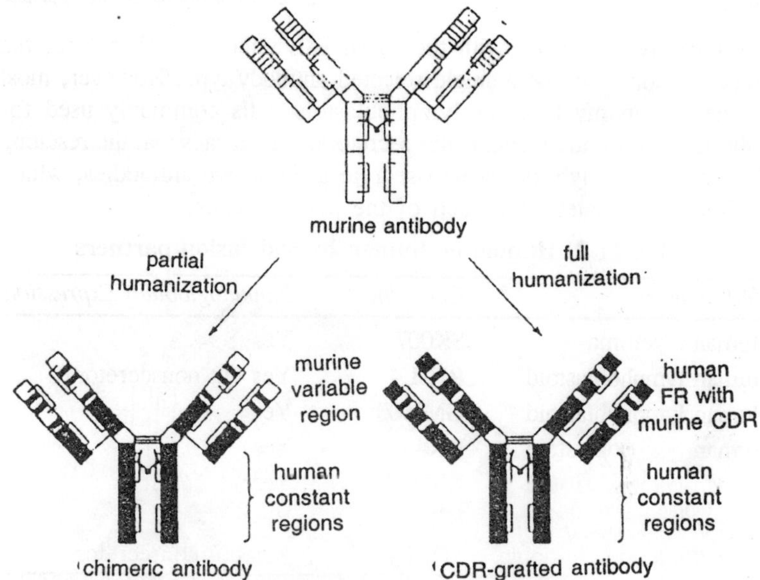

Fig. 11.8. Ways of humanize an antibody.

in humanized therapeutic antibodies have allowed the introduction of a range of products against specific human diseases.

Recombinant Antibody Fragments

Various fragments of human immunoglobulins have been expressed successfully in bacterial cells. These include the Fv fragment, the single-chain Fv fragment (scFv), the Fab fragment and the F(ab)2 fragment. The Fv is the smallest antigen-binding fragment of an immunoglobulin with a molecular mass of around 25 kDa. The VH and VL domains of the Fv fragment are stabilized by disulfide bridges. In the scFv fragment a short peptide spacer (usually 15-20 amino acids) is introduced in order to link the VH and VL domains covalently. This also allows the possibility of the linkage of two scFv fragments to create 'diabodies' which are bispecific in so far as they have two independent antigen-binding sites.

Bispecific antibodies can also be produced from the fusion of two hybridomas to generate a 'quadroma'. However, all combinations of light and heavy chains are synthesized in these cells with only a few of the molecules being bispecific. Purification of the required molecules would be a difficult task. The potential advantage of these recombinant fragments for human therapy is their small size that facilitates tissue

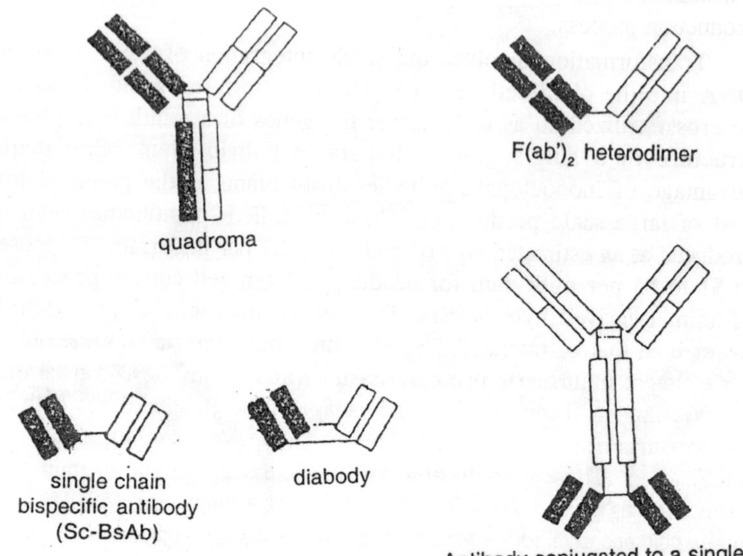

Fig. 11.9. Recombinant antibody fragments.

penetration, biodistribution and blood clearance. The fragments can be isolated from libraries of antibodies displayed on the surface of filamentous bacteriophages. This phage display technology is an alternative strategy that can be used instead of mammalian hybridoma technology. The disadvantage is that the recombinant antibody fragments lack glycosylation and also the binding sites for complement and FMC receptors. However, the possibilities exist of conjugating other. polypeptide sequences to express the desired effector functions. Conjugation of toxins or specific growth factors to these fragments also allows the future development of immuno-constructs with considerable potential for therapeutic activity.

Potential Alternative Methods of Production

Antibodies from Plants

Antibodies were first expressed in transgenic plants in 1989. Since then various antibody fragments and domains have been produced in plant hosts as well as full-length and multimeric antibodies. The most popular host species for this work has been the tobacco plant, *Nicotiana*, although corn and soybeans have also been utilized. There is no apparent reason why other plants could not be used. The value of using plants for monoclonal antibody production includes the absence of animal pathogens, the ease of genetic manipulation, the ability of post-translational modification and the potential for scale-tip to an economic production process.

Transformation involves the stable integration of the appropriate DNA into the plant cell genome. The resulting transgenic plants can be crossfertilized so as to integrate the genes of a multimeric protein structure in a single genetically stable hybrid plant. One major advantage of monoclonal antibodies from plants is the potential low cost of large-scale production. These so called 'plantibodies' can be produced at an estimated cost of $0.01 to $0.1 per milligram as opposed to $1 to $5 per milligram for production from cell culture processing of animal-derived hybridomas. The cost of microbial fermentation is lower than that of mammalian cell culture but bacteria lack the ability for efficient multimeric protein assembly and of any post-translational modification. A further potential advantage of the plantibodies is delivery by consumption of plant tissue and thus avoiding any need for purification. These possibilities are particularly applicable in certain cases such as the previously shown ability of a plant-produced antibody against *Streptococcus mutans* to prevent binding of the bacteria to the surface of teeth, thus reducing tooth decay.

Plant cells are eukaryotes and therefore capable of post-translational modification of proteins including N-linked glycosylation. However, although the plant glycan structures have not been analyzed in detail it is likely that these structures are significantly different from those in mammalian systems. For example, the commonly found mammalian terminal sialic acid (N-acetyl neuraminic acid) residue is a structure not found in, plants. Also the α-1,3 core fucose structure appears to be unique to plants and has been implicated in human allergies to pollen. The potential presence of such unusual glycan structures in plant-derived antibodies might not have an effect on antigen binding but for a therapeutic antibody they are likely to increase the chance of an adverse immunogenic reaction during human treatment. This could limit the use of plant-derived antibodies in certain applications particularly if systemic long-term administration is required.

Humanized Antibodies from Transgenic Mice

Transgenic mice strains have been produced capable of synthesizing human monoclonal antibodies. Xenomouse strains have large portions of human variable region genes incorporated into the germ line via a yeast artificial chromosome (YAC). The megabase-sized YAC allows the genes for human heavy and light chain immunoglobulin to be incorporated as transgenes into a mouse strain deficient in the production of murine Ig. The large human variable region repertoire incorporated as transgenes allows the mice to generate a diverse immune response comparable to that in humans. These human genes are also compatible with the mouse enzymes that allow class switching from IgM to IgG. The immunoglobulin generation will also undergo somatic hypermutation and affinity maturation, a natural process that enhances the affinity of the antibody for the antigen. Thus an antigen introduced into a Xenomouse produces a human monoclonal antibody with high specificity for its corresponding antigen. Such antibodies have already proved their potential in clinical trials.

Importance of Glycosylation to Therapeutic Antibodies

The glycosylation pattern of the immunoglobulin structure has particular relevance to the production and use of monoclonal antibodies as therapeutic agents. Any *in vitro* production process results in a heterogeneity of glycan structures of the product protein. To avoid any undesirable immune response in the use of such antibodies it is important to maximize the content of fully processed glycans. There are various parameters that affect the glycan processing from the metabolic profile of the hybridoma to the environmental conditions of culture. It has been shown that the glycan structures vary with the specific activity of key glycosylating enzymes contained in a hybridoma. This in turn depends upon the enzymic profile of the parental cell lines used in the hybridization process. In one study hybridomas were produced from parental cell lines, only one of which had an enzyme (UDP-N-acetylglucosamine: β-D-mannoside β-1,4-N-acetylglucosaminyl-transferase; GnT III) responsible for the addition of a bisecting GlcNAc. As expected the resulting hybridomas had varying levels of GnT III. Of interest was the fact that the content of bisecting GlcNAc in the antibodies produced by each hybridoma was a reflection of the intracellular activity of GnT III.

A further example of the effect of the producer cell line on the glycosylation pattern of a monoclonal antibody has been shown for the IgG, CAMPATH-1H. This is a humanized recombinant murine monoclonal antibody developed for human therapy and has been expressed in various cell lines including a murine myeloma and Chinese hamster ovary (CHO). The glycosylation of the antibody produced from CHO was consistent with normal human IgG. That is fucosylated biantennary structures containing zero, one or two galactose residues. However, the immunoglobulin from a murine myeloma (NSO) results in some potentially immunogenic glycoforms containing Galα(1-3) Gal terminal residues. Such hypergalactosylated proteins have been shown following expression from various murine cells.

Cell culture conditions have also been shown to affect product glycosylation. Relevant culture parameters include the accumulation of ammonia, the dissolved oxygen level, glucose depletion, lipid composition, pH and protein content of the medium. The glycosylation of monoclonal antibodies from hybridomas is particularly susceptible to the dissolved oxygen level of the culture, which on a large scale is often maintained at a specific setpoint through an oxygen probe that controls the gaseous input to the bioreactor. This dissolved oxygen (DO) is usually calibrated from 0 to 100%, which is the level of oxygen relative to saturation with air. The relative proportion of three predominant glycoforms changes depending upon culture conditions. For example, the dissolved oxygen level of a hybridoma culture has a noticeable effect on this distribution. Whereas a normal distribution of G0, G1 and G2 is found at 50% DO, lower levels of oxygen (<10%) may lead to poor galactosylation and a consequent increased proportion of G0.

12

Preservation of Animal Cell Lines

Thousands of cell lines derived from human and animal sources are in existence today. Some derived from normal tissues have a limited growth potential, whilst others, which have been derived from tumours, or transformed to behave as tumour cells, can be propagated in continuous culture for many years.

As a cell line evolves from primary culture to form a secondary or established cell line there is a danger of introducing genetic or phenotypic instability or contamination with other cells, bacteria, fungi or mycoplasma.

It is essential to use cell lines which are well defined and free from contamination in order for experimental results to be valid. For this reason it is vital that laboratories using animal cell lines should:

1. Standardize culture conditions;
2. Select a period in the life of the cell line where variation is at a minimum;
3. Select a pure clone, or well characterized cell line, if possible;
4. Preserve a seed stock of the cell line which can be used to replenish the working stock at intervals.

Maintaining cultures simply by sub-culturing is expensive in terms of time and materials. This system is also susceptible to contamination. Every time we open a vessel to inoculate it, there is a finite chance of contaminant micro-organisms gaining access to the culture. These problems are particularly acute when cultures are needed to be maintained for use over many years. This chapter examines the techniques available for long term storage of cell lines.

Variation and Instability in Cell Lines

Culture Conditions

Once the optimal conditions for maintaining a cell line have been established, these should be adhered to as far as possible. Any changes in the constituents of the cell culture medium, batch differences in serum, differences in temperature and gases can alter the growth characteristics of the cell line. Laboratories which regularly work with animal cell lines usually buy *batch-tested medium* and serum in large quantities often sufficient to last around a year.

Selective Overgrowth can Rapidly Alter a Cell Line

As noted already that in a primary culture there is a number of different cell types growing at different rates. Some may grow very quickly and overgrow the more slowly dividing or non-dividing cells. For example, if cells are explanted from a colon carcinoma the first culture will contain cells which are mainly epithelial. After sub-culture, however, these epithelial cells diminish in number and the fibroblast cells take over. Eventually all the epithelial cells disappear and the fibroblasts form a secondary cell line.

If, for a particular experiment, the epithelial cells need to be preserved then it will be necessary to produce a clone of the cells required and also to freeze down some of the primary culture in case disaster strikes and a clone cannot be produced. Once the required clone is obtained, it is vital to freeze down a seed stock which can be used to replenish the working stocks should these get contaminated or lose the particular characteristics required.

A Transformed Cell can Appear in an Established Cell Line

Sometimes a cell can become transformed during the life history of an established or semi-established cell line and take over very rapidly. The transformation event may enable it to have a much greater growth capacity than the slower growing parent cell line.

Genetic Instability can Alter the Phenotype of a Cell Line

Established cell lines have a high incidence of genetic variations partly because the rate of spontaneous mutations is higher in cells which are proliferating rapidly and partly also because there is no mechanism *in vitro* for the elimination of mutants as they appear. These genetic changes can be expressed as phenotype changes and thus the characteristics of the cell line may change altogether. Cells which are to be preserved should be genetically characterized.

Senescence

A finite fibroblastoid cell line such as that derived from human embryo lungs will be stable for only a limited number of generations. The cell line will become stable at around the fifth or sixth passage (generation) and can be used up to the 30th generation. After this senescence can become a problem and cells may lose certain receptors as well as the ability to replicate.

By preserving some of the early sub-cultures, it is possible to extend the time in which work can be conducted on the cell line. This situation can be represented in the following way.

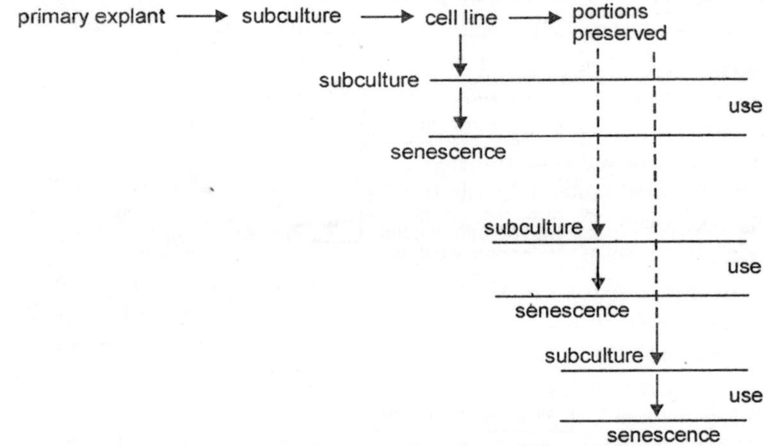

Preservation of Cell Lines

Historical Background

In 1776 Spallanzi published one of the earliest observations of the effect of' cold' on cells. He noticed that spermatozoa chilled with snow become motionless and that this loss of activity was totally reversed upon increase in temperature. A hundred years later Mantegazza (1866) experimented on the preservation of spermatozoa and noted that cells show a good degree of resistance to freezing and thawing. This work had relevance in animal husbandry, but it was not until the 1940s that work on the preservation of cells other than spermatozoa really took off. In 1945 Parkes showed that the survival of stored cells was increased by the presence of glycerol and by increasing the cell density. Around this time other workers noted that the major cause of cell damage could be attributed to the rise in concentration of electrolytes as water was removed into ice crystals.

The preservation of animal cells has become dependent upon using low temperature to render cells metabolically inert. The term *cryopreservation* is given to this method of cell storage. As we shall see, for optimum results storage is usually done at liquid nitrogen temperatures (-196°C) although temperatures of -70°C to -80°C may be sufficient to maintain cell viability for a few months.

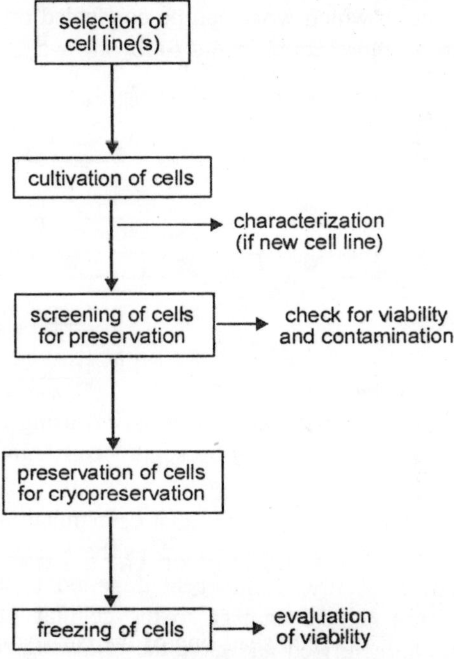

Fig. 12.1. Stages in the cryopreservation of animal cells.

Cells preserved in the late log phase are more resistant to the traumas of freezing and thawing and will continue to replicate once thawed. It is usual to test the viability of cells before cryopreservation. This can be done using a staining procedure which enables a distinction to be made between viable and dead cells. Trypan blue is commonly used. It will stain dead cells but is excluded from live cells. Usually viabilities greater than 95% are required and cultures showing lower viabilities are usually discarded. Cells to be used for storage are collected using the same methods as used for their routine sub-culture.

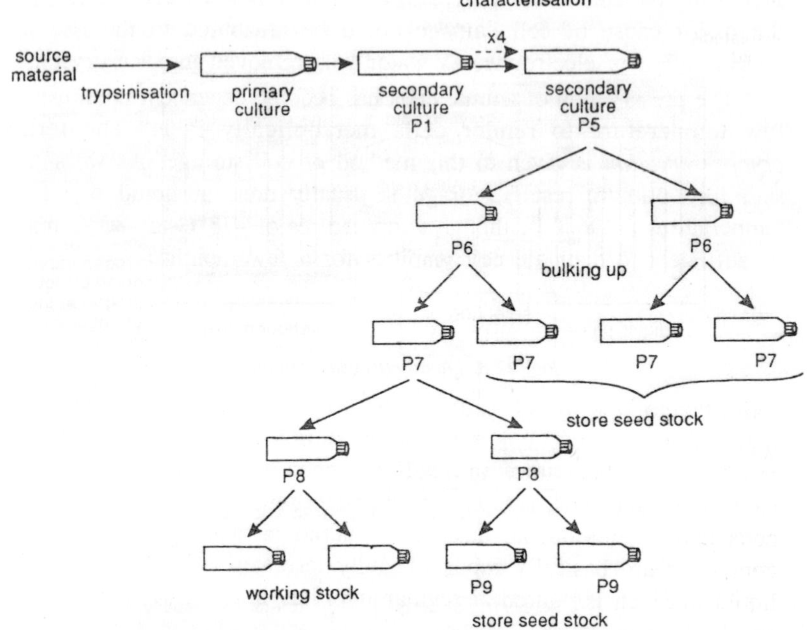

Fig. 12.2. Selecting cells for storage. P refers to the number of passages after primary cell culture.

For example, attached cells are removed using trypsin whilst cells in suspension may be collected by gentle centrifugation. Collection must, of course, be done aseptically.

The two main reasons are that centrifugation reduces the volume that needs to be frozen which enables the freezing process to be more controlled. Secondly, gentle centrifugation is used to minimize the extent of mechanical damage to the cellular membranes. A culture has been characterized and gave the following results.

1. Cell density = 1×10^6 ml^{-1}
2. Free of bacteria.
3. Free of mycoplasma.
4. Free of fungi
5. 85% viable (as tested using trypan blue).

Freezing Down

Temperature of Storage

Cells can be stored for varying lengths of time at temperatures between -70°C to -196°C. For short term storage, for example for no

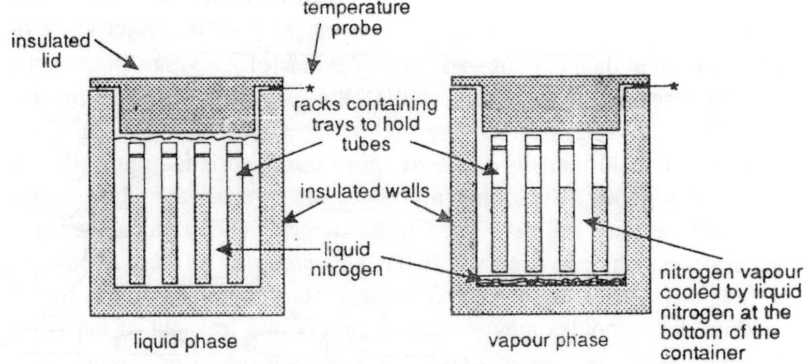

Fig. 12.3. Liquid nitrogen containers.

more than a few weeks, cells can be preserved at –70°C in standard mechanical freezers. For longer term storage, it is necessary to use much lower temperatures and liquid nitrogen is the medium of choice. Cell banks and most reasonably sized laboratories routinely preserve cells and other valuable material in liquid nitrogen containers. The containers are basically robust, heavily insulated, vessels into which liquid nitrogen is poured at regular intervals in order to maintain the required temperature.

Cells can be stored in either the vapour phase (–120°C) or the liquid phase (–196°C) of liquid nitrogen. Storage at –120°C does not significantly reduce cell viability, although for very precious material or for indefinite storage the liquid phase is often preferred. There are advantages and disadvantages associated with both systems. Storage in the vapour phase is generally safer: seepage of liquid nitrogen into the storage tubes (*ampoules* or *cryotubes*) does not occur so the risk of explosion is reduced. If liquid nitrogen enters the ampoule, the sudden evaporation of the liquid on removal of the ampoule from the cryostat may cause the ampoule to explode violently. The risk of the contents of the tubes, which may contain infectious material, leaking out of the tube and contaminating the liquid nitrogen and, thus, other tubes is also reduced. For most purposes the vapour phase provides the ideal environment for storing cells.

Liquid phase temperatures (–196°C) are advisable when the material is to be stored for very long periods, often longer than the life of the scientist! Particularly precious material should also be stored in the liquid phase so that if a leak develops in the freezer and liquid nitrogen evaporates there is still a time delay before the material is lost. Many –70°C freezers and liquid nitrogen containers are fitted with computerised alarms so that engineers can be called out automatically at any time should the temperature begin to rise to dangerous levels.

The main dangers come from explosions of ampoules or cryotubes. For this reason it is essential that the operator wears appropriate protective clothing (protective goggles, face mask, gloves). Injudicious handling of cold utensils can also cause burns so it is important to use handling tongues. Exploding ampoules also means that the preserved cell culture may be lost. Thus it is imperative to reduce the risk of such losses. One way is to transfer the *ampoules* to the gaseous phase a few days prior to use. In this way, any liquid nitrogen that has entered the ampoules may evaporate slowly and leak out of the vessels.

The cells are stored, ideally at a concentration of around 10^6-10^8 cells per ml in tubes designed specifically for very low temperatures. These cryotubes are made of resilient polypropylene and have a rubber 'O' ring in the screw caps to prevent leakages which could lead to explosions upon thawing. Alternatively, glass ampoules may be used. These, however, need heat-sealing apparatus and considerable skill to ensure that they are sealed properly. They are also quite difficult to open whilst ensuring asceptic conditions are maintained. It is not surprising, therefore, that the screw-topped plastic cryotubes are preferred by most workers.

Preparing Cells for Freezing

As discussed already that cells for freezing are prepared by trypsinization to detach adherent cells from flask surfaces. Cells in suspension are centrifuged at around $100 \times g$ for 5-10 minutes and the pellet is resuspended in a small volume (1-2 ml) of storage medium. The composition of the medium used to suspend cells is rather specialized. It contains a cryoprotectant (usually glycerol or dimethyl sulphoxide (DMSO)) and a high protein concentration (usually extra serum). The formulation of this storage medium has evolved through many trial and error experiments. The usual concentrations of additives are;

- cryoprotectants 7-10% (v/v);
- serum 20% v /v.

If DMSO is used, care must be taken to avoid direct contact as it is a powerful solvent that can penetrate skin. It is also readily oxidized and the products of oxidation are toxic to cells. Thus fresh DMSO should be used and ideally it should be stored at –20°C.

Freezing is Stressful

When viable cells and tissues are frozen a number of stresses are set up within the intracellular environment. At around 0°C ice crystals begin to form and these crystals grow as the temperature drops. Because the crystals are made of water alone, the result is that the constituents of the cell which are normally dissolved in water become more and more concentrated as the water freezes. This process of *dehydration* creates osmotic pressures within the cell which are thought to be the major cause of cell damage and death.

The formation of ice crystals is only a problem, however, when the drop in temperature is rapid. A great deal of the damage can be prevented if the rate of cooling from 0°C to 50°C is slow and controlled. A fall of about 1°C per minute is optimal. Once a temperature of -50° C is reached cells must then be cooled rapidly to the final holding temperature.

Rapid chilling results in thermal shock and leads to cell death or injury. Some mechanical injuries to cell membranes during freezing also occur due to ice crystals forming in the extracellular matrix. This mechanism is not thought to be the most common cause of cell death and can be avoided by using a cryoprotective such as DMSO at 7-10% final concentration in the storage medium.

Slow Freezing or Controlled Freezing is Vital

The cells suspended in a small volume (0.5-1.0 ml) of storage medium should be carefully labelled identifying the cell type, approximate cell density and date of storage. There is usually little more room to put any other information on the actual tube but detailed records should be kept on index cards or computer databases.

Sophisticated devices are available to control the freezing of cells at a rate of 1°C per minute. Programmable freezers are useful in laboratories whose main purpose is cell storage but are generally too expensive for the smaller laboratory. A crude but effective alternative is to place the tubes to be frozen in an upright position in a polystyrene box which can then be placed at –70°C overnight or until the cells have frozen to below -50°C. They can then be transferred immediately to liquid nitrogen for rapid cooling to –120°C or –196°C.

Keeping Records and Labelling

It is vitally important to keep proper records of the preserved cell lines. Remember that these cultures may be kept for a considerable length of time. In many cases, the preserved cultures may be kept long after the original depositor has left. The information that needs to be kept is quite extensive. Here we will outline the main features. We have put this information under a number of headings.

- Features of the cell line: origin, cell type, cell line number, tests carried out, for example for viruses, mycoplasmas etc;
- Date of preservation/ storage;
- Person responsible for deposition;
- Location in the store;
- Cell density (approximate);
- Number of ampoules/tubes stored;
- Number and date of removal of ampoules, details of where the culture has been used results of viability checks;
- Other special features (for example if cited in a patented process).

This type of information is kept on index cards or, increasingly, is stored electronically. You may well imagine that for cell banks (see below) that preserve many thousands of cell lines, keeping records is a major and integral part of cell line preservation. It is essential that when the number of ampoules is reduced to a particular critical level, the stock is replenished by repeating the cultivation-preservation cycle.

The number of ampoules produced and the regularity of re-furbishing these stocks is, of course, governed by the anticipated and actual use of the stocks. It is also important to know who has used stock cultures. If a subsequent problem arises (for example if the culture is found to be erronously labelled or contaminated with a virus) the recipient can be traced and alerted.

Thawing Frozen Cells

Because of the problems of genetic instability and contamination mentioned earlier, it is necessary to replace the working stocks from time to time. Cells retrieved from storage must be thawed rapidly to ensure maximum survival. Where possible the ampoule should be plunged into a beaker of water at 37^{0}C until thawed. A face mask or visor and gloves should always be worn to protect skin and eyes from the liquid nitrogen and also for protection in case the ampoule explodes as the liquid nitrogen expands.

As soon as the cells have thawed completely (around 30 seconds) the cell suspension can be transferred drop wise into a centrifuge tube or container containing about 20 ml of pre-warmed growth medium supplemented with 10% foetal calf serum.

The main reason of adding the cells dropwise to the medium is that this approach rapidly dilutes the cryoprotectant. In many cases, the rate of removal of the cryoprotectant is critical to the survival of the cells. Some cells are particularly sensitive to cryoprotectants and it may be necessary to remove the cryoprotectant by immediate centrifugation. The washed cells can then be transferred to a small 25 cm^2 flask containing fresh growth medium to recover for a few days.

The cell density used during resusication is important. If too Iowa density is used, the cells remain in a lag phase for a very long time. Usually a density of 2-5 $\times$ 10^4 cells ml^{-1} is recommended for anchorage-dependent cells and 3-6 $\times$ 10^5 cells ml^{-1} is recommended for anchorage-independent cells.

QUANTITATION OF CELL VIABILITY

It is useful to quantitate the rate of recovery following freezing for two important reasons:

1. Quantitative assessment of cell viability will give some idea of how soon the cells will recover sufficiently to be used or passaged;
2. A low number of viable cells may indicate problems with the storage protocol or temperature failure during storage.

Dye Exclusion Test

This test is crude but gives a rough estimate of cell viability very quickly—so valuable time and reagents are not wasted trying to maintain cells which are already dead. As we have seen in this technique, an isotonic solution of a dye such as trypan blue (0.025%) or erythrocin B is added to a small aliquot of the cell suspension. Cells which have retained membrane integrity will not have taken up dye, whereas those whose membranes have been damaged during the freezing and thawing will have had the dye diffused through into the cytoplasm. Cells which do not take up the dye can be counted using a haemocytometer.

Clone-forming Efficiency Test

This is a more accurate test and it can be used for adherent cells. An aliquot of the reconstituted cells is diluted to give between 10^2 and 10^4 cells ml^{-1}. Serial tenfold dilutions of this cell suspension are made in growth medium and each dilution is inoculated into small 25 cm^2 flasks. The flasks are incubated at 31°C for 12-14 days after which the cells are fixed with formaldehyde. The fixed cells can then be stained with a simple dye (e g. 1% toluidine blue) and examined under the microscope.

The clone-forming efficiency is calculated as:

$$\frac{\text{no of clones formed}}{\text{no of cells inoculated}} \times 100\%$$

The clone-forming efficiency test gives a measure of the number of cells which actually grow. In contrast, the dye exclusion test merely measures the number of cells which do not take up dye. It is by no means certain that all such cells are capably of growing and dividing.

The advantages of the dye exclusion test are that it is simple, requires little specialist equipment and gives a rapid result. The clone-forming efficiency test, however, requires substantial media preparation and the results are only obtained after 12-14 days.

Cell Banks

There are a large number of culture collections which maintain large stocks of animal cells. These collections fulfil a number of functions. Most importantly they provide resource centres from which cell lines may be obtained. 'This is a two way process. Scientists can submit the cell line they have obtained from their own research. Alternatively they can obtain cultures for research or teaching purposes.

The culture collections also provide a variety of other services. For example they will characterize cell lines and, in some instances, provide training resources for the maintenance and characterization of

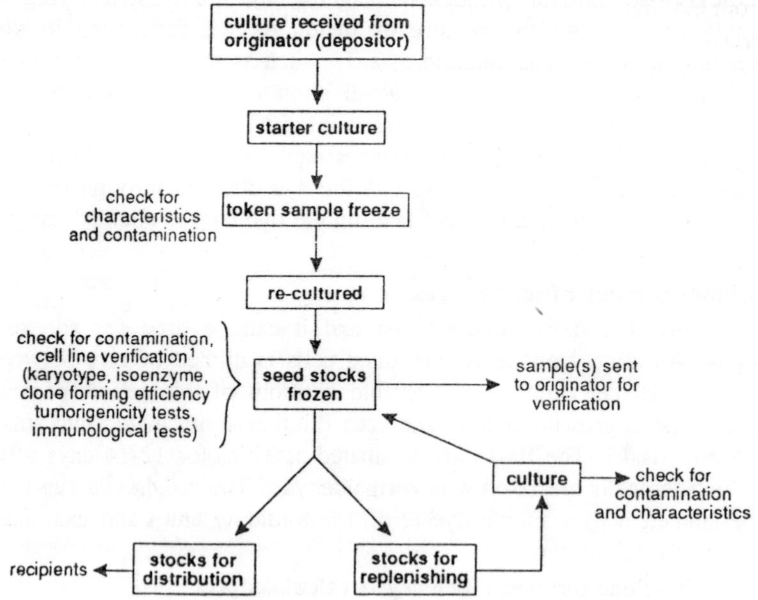

Fig. 12.4 Schematic representation of the procedure used to verify and maintain cell lines in cell culture collections.

cell lines. They also act as depositions for patented cell lines. We will not go into details of the patenting process here, except to say that patenting procedures demand that relevant cell lines are submitted to a recognized cell culture organization.

You should realize that a number of conditions are attached to the deposition and acquisition of cell lines. Although the major aim of culture collections is to make cultures available to the scientific community, the conditions of release are governed by concerns of safety and the commercial value of the cell lines. Patented organisms are not generally released. Likewise the release of cells which present some clinical danger are usually restricted to those who can demonstrate that they have appropriate containment faculties and expertise.

Almost all culture collection services have to offset some of their operational costs by charging for the supply of cultures and they all operate within the legal framework for transporting cultures. This includes import and export requirements. Also of concern is the need to ensure that cultures remain viable during shipment.

In all culture collections, a key question is quality assurance to ensure that cultures are not contaminated and that cultures received from external sources are exactly as the depositor believe them to be. Equally, recipients of a culture need to be assured that the culture they receive is properly accredited.

13

Establishing Cell Products

Animal cells synthesize a vast range of compounds that can be extracted and purified from cultures. Many of these compounds have potential value as healthier products. The initial challenge for the clinical application of these products is to isolate a sufficient quantity at an acceptable level of purity. There have been three major innovations in the development of mammalian cell cultures for the production of biologicals in the last few decades. Each has led to the production of a new category of biologicals.

1. *Viral vaccines*: from the ability to grow viruses in cell culture.
2. *Monoclonal antibodies*: from the ability to generate hybridomas by cell fusion.
3. *Recombinant proteins*: from the ability to transfect cells with isolated genes and amplify to allow high level of expression of the corresponding proteins.

The use of these products in human clinical applications will be discussed in this chapter. Expansion of specific cell lines has been used in such areas as toxicity testing and various clinical procedures.

Viral Vaccines

Enders discovered in 1949 that the poliomyelitis virus could be grown from primary monkey cells in culture. The virus has an icosahedron structure with an inner core of nucleic acid and an outer coat or capsid which is made of repeated protein capsomere structures. The polio vaccine, produced in 1954, was the first human vaccine to be produced using large-scale cell culture techniques. This eventually led to the development of animal cell technology for the production of a range of human and veterinary viral vaccines against a variety of diseases. This offers a more consistent and efficient method of production compared to alternatives such as the propagation of virus in chick embryos, a method that has been used even up to recent times.

Table 13.1. Examples of virus vaccines produced in large quantities

Human	*Veterinary*
Polio	Foot-and-mouth disease
Measles	Marek's disease
Mumps	Newcastle disease
Rubella	Rinderpest
Yellow fever	Rabies
Rabies	Canine distemper
Influenza	Swine fever
	Blue tongue
	Fowl pox

Childhood vaccination is now routine and the importance of these vaccines in the improvement of world health over the last half century has been enormous. Polio vaccines exist as inactivated viruses (Salk) or live attenuated viruses (Sabin). Both have been used extensively throughout the world with the result that incidence of poliomyelitis is now extremely low in the West. Smallpox was completely eradicated as a world disease in the 1960s and now has only arisen as a potential weapon of bioterrorism from the culture stocks that were retained in specific designated laboratories. Both of these diseases were at one time endemic and widespread, particularly among the young.

The basis of vaccination is to inject a viral antigen in a nonpathogenic form. This induces the formation of a corresponding antibody which can provide some protection against the live virus and the associated disease. The two major vaccine types are listed below.

1. An inactivated pathogenic virus. This is a disease-causing virus which has been chemically inactivated.
2. An attenuated live virus. This is capable of propagation but has been changed genetically so that it can not produce a disease. However, the virus still has the surface proteins capable of eliciting an immune response which protects against the pathogenic virus. Live viral vaccines are often preferred because they can be administered in small but effective doses.

Production of Viruses

Animal viruses can be propagated by inoculation into a culture of cells at relatively high density. The cell type should be susceptible to the particular virus required. Anchorage-dependent cell growth is particularly well suited for viral growth and for large-scale production, with microcarrier cultures often being used. The lytic cycle consists of four distinct phases of adsorption, penetration, replication and release.

During this cycle, cellular metabolism is directed to the formation of new virus particles and eventual cell lysis. However, some viruses do not cause cell lysis but arise by budding from the cell membrane. The quantity of virus is usually expressed in plaque-forming units (pfu) which relates to the formation of a plaque on a monolayer of cells. The plaque is recognized under a microscope by a clear spot in the monolayer caused by cell lysis as a result of infection by a single virus. A plaque assay involves addition of a virus suspension at various dilutions to cell monolayers in a series of Petri dishes.

The pfu values are determined from counts on the Petri dishes in which distinct and separate plaques can be distinguished. For production

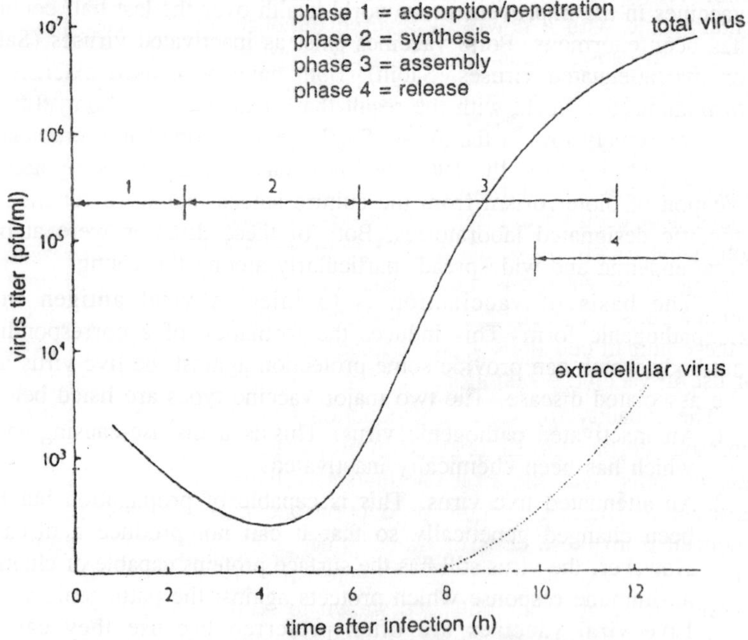

Fig. 13.1. Phases of viral growth in cell culture.

the virus is added to a cell culture at a multiplicity of infection (moi) of 0.1-10 pfu/cell with the expectation that this will increase to 10^3-10^4 pfu/cell following a period of propagation. This represents an increase of up to $\times\ 10^5$ of the total virus in the culture. It is important to identify the time of maximum titer of extracellular virus. This normally occurs about 12-24 hours after viral inoculation and is the time at which the viruses should be harvested and purified from the culture. The outline given here is intended only as a brief guide to the events associated with viral infection and production.

Cell Lines for Vaccine Production

Safety is of major concern when choosing a cell line for vaccine production. During the development of vaccine production processes, it was decided that the perceived risk of using tumourigenic cell lines such as the well-characterized HeLa cells as a substrate for viral propagation was unacceptable. In 1954 the regulatory bodies in the USA decreed that human vaccines should be produced from primary cells derived from healthy monkeys. The poliomyelitis vaccine was subsequently produced by viral propagation on primary monkey kidney cells. However, it was discovered later that these cells were contaminated with a newly discovered tumourigenic virus designated SV40. This caused concern even though there was no evidence for the transmission of the SV40 virus to recipients of the vaccine.

Extensive study and characterization of normal human diploid fibroblasts in the early 1960s led to their acceptance for vaccine production. Two human lung fibroblast lines named WI-38 and MRC-5 have now been used extensively for polio vaccine production. Even though these cells have a finite lifespan, their use in the large-scale propagation of viruses is much easier than the alternative of obtaining primary cells for each batch of vaccines. Consequently, large stocks of these human cell lines at low passage number have been maintained for use in vaccine production.

The capacity for vaccine production from an individual strain is enormous despite the finite limit on propagation. Vero (African green monkey cells) was the first continuous cell line to be accepted as a substrate for human vaccine production and polio vaccine has been produced from these cells in France since 1983. This is a continuous cell line with excellent growth characteristics. Despite its heteroploid genetic character it has not been shown to be tumourigenic.

A live virulent virus requires complete inactivation before it can be used as a vaccine. Treatment with formaldehyde can be effective providing the treatment is for a sufficient time to ensure inactivation of all virus particles. An alternative vaccine is the live attenuated virus. This has the antigenic determinants of the normal virus but has been modified genetically so that it is no longer pathogenic. The attenuation can be induced by mutagenic agents but in most cases it occurs spontaneously by continuous passage of the virus, in cell culture. Vaccines can also be produced as peptides which mimic the antigenic effect of the viral coat protein.

Monoclonal Antibodies

All useful applications of monoclonal antibodies depend upon their high specificity of binding to an antigen. Many Mabs produced commercially are used for diagnostic tests for the identification of small quantities of specific antigens. Changes in the level of hormones or enzymes in blood or urine can be used to check for pregnancy or any change of. physiological state. Immunodiagnostic tests of this type can be performed within 15-20-min, which has considerable benefit for the early diagnosis and in the treatment of disease.

Therapeutic Antibodies

Mabs also have applications as therapeutic agents, particularly in targeted drug therapy. In 1986 the monoclonal antibody named OKT3 developed by Ortho Pharmaceuticals was the first murine Mab to be licensed for therapeutic use. OKT3 recognizes a surface antigen (CD3) on T-lymphocytes and is one of the most effective agents in preventing immunological rejection of transplanted kidneys. Of course, therapeutic Mabs take a longer time to develop because of the need for the strict clinical trials that are required before any injectable compound can be licensed for human treatment. Clinical trials that demonstrate safety are often the most costly part of developing a new medicinal product.

Various monoclonal antibodies have been targeted to membrane-bound protein antigens specifically expressed in tumour cells. These antibodies can be designed in configurations likely to cause the destruction of the target tumour cells. The conjugation of radioactive or toxic compounds to the antibody can result in a localized high concentration resulting in cytotoxicity to the target cells. Alternatively, an enzyme may be conjugated that will catalyze the release of a toxic product from an ingested 'pro-drug'. Another strategy is to promote an effector function through the use of a bispecific antibody that could potentially activate T-cells leading to the specific destruction of the target cells.

The rationale behind these methods is to cause localized cell destruction but limit systemic toxicity. However, the results of the initial clinical trials for therapeutic murine antibodies using these strategies were disappointing. This was as a result of unexpected toxicity associated with treatment of immunoconjugates and the undesirable human anti-murine antibody (HAMA) immune response. However, the administration of antibodies relying on a response within a short period of time (up to 10 days) was more successful. This included antibodies for radioimaging, radioimmunotherapy or for acute allograft rejection (the OKT3 antibody).

The development of human chimeric antibodies in the 1990s increased rapidly the rate of licensing of monoclonal antibodies as therapeutic agents. These humanized antibodies do not elicit the HAMA effect and also the half-life is much longer than mouse antibodies. Most mouse IgGs have a half life of less than 20 h whereas an antibody with a human-type constant region can have a half-life of up to 21 days. As well as these, there are many more antibodies in clinical trials with the expectation that the numbers in therapeutic use will increase in the future.

The success of these chimeric antibodies can be illustrated by Rituxan which is used in the treatment of non-Hodgkin's lymphoma. This has a murine variable region which binds specifically to CD20 on B-cells and a human Fc domain to trigger effector mechanisms. CD20 is a, protein expressed by over 90% of the lymphoma cells. These tumour cells can become coated by the anti-CD20. This results in activation of the complement pathway and Fc receptor-bearing cells which can destroy the tumour cell.

Therapeutic Recombinant Glycoproteins from Mammalian Cells

Proteins extracted from biological sources have been important for substitution therapy since the 1920s when Best and Banting used insulin to treat diabetes. However, the problems associated with such extraction processes include the large quantity of tissue, blood or urine required for extraction of the small amount of each product. In some cases the problem could be solved by extraction from nonhuman sources, although this always poses the risk of an adverse immune response in human recipients. In the case of insulin this was solved by the chemical transformation of pig insulin which differs from the human version only by one amino acid. However, the overriding reason for abandoning most of these extraction processes is the risk of contamination from

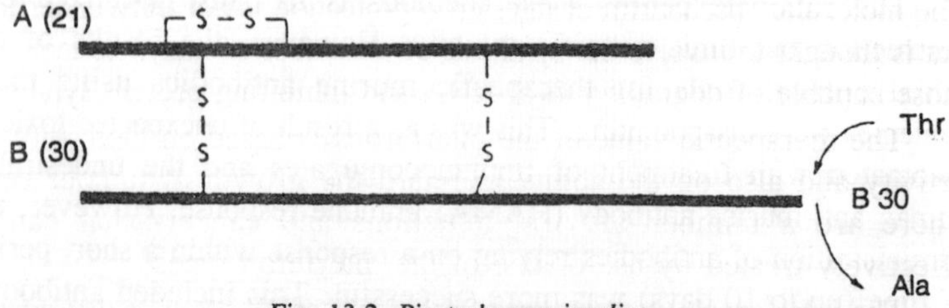

Fig. 13.2. Pig to human insulin.

infectious agents such as HIV and prions. In most cases these are natural products that can be produced in large quantities following the isolation of a specific gene which is then expressed in a host cell line.

Interferons

In 1957 Isaacs and Lindenmann discovered a compound that 'interfered' with viral propagation in cells in culture and introduced the term interferon. They showed that culture medium taken from cells that had supported viral growth could protect noninfected cells from a subsequent viral infection. This led to the identification of a group of proteins that were shown to interfere with the propagation of viruses and subsequently named 'interferons'.

An interferon refers to an inducible secretory protein produced by eukaryotic cells in response to viral and other stimuli: The Interferons are now recognized as a family of cytokines having antiviral, antiproliferative and immunoregulatory effects. Viral infection induces the synthesis of interferon by derepression of the gene. The secreted interferon can then induce an 'antiviral state' in neighbouring cells.

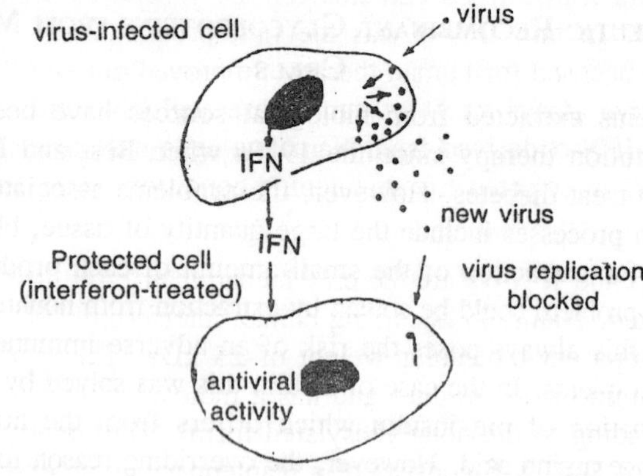

Fig. 13.3. Interferon 'interferes' with viral replication in protected cells.

The molecular mechanism of interferon protection is not fully understood but is thought to involve the synthesis of a number of enzymes including those capable of degrading viral RNA and inhibiting protein synthesis.

The therapeutic value of the interferons is based on their antiviral activity and also on the ability to retard the growth of tumour cells. There are a number of viral infections and cancers that can be effectively treated by use of one of the interferon types.

The interferon's are a group of relatively small proteins that contain 140-170 amino acids and are divided into three main types—IFN-α, IFN-β and IFN-γ. A further complexity arises in that there are at least-25 independent subtypes of IFN-α which are all products of independent genes. Each interferon type is derived and secreted predominantly by one cell type—B-lymphocytes (IFN-α), fibroblasts (lFN-β) and T-lymphocytes (IFN-γ)—and this accounts for differences in production processes.

IFN-α

A procedure for the isolation of IFN-α from leukocytes obtained from pooled human blood was pioneered in the 1960s but the yield from this process was too low to satisfy the growing demand for interferon. Subsequently a large-scale cell culture production process was developed by Wellcome (UK) to operate in 10,000-liter batch fermenters. This process is based on a human B-lymphoblastoid cell line (Namalwa) that is a particularly good producer of IFN-α following induction with Sendai virus.

The cells grow well in suspension and have been adapted to low-cost serum-free medium. This is one of the few examples of the use of a human transformed cell line for the production of a therapeutic compound. In fact IFN-α was the first product from a tumour cell line to be licensed for human therapy. Improved purification techniques and sensitive assays for DNA can ensure negligible contamination. On this basis IFN-α derived from Namalwa cells has gained approval for routine use in selected life-threatening conditions (e.g. leukemia).

IFN-β

IFN-β and IFN-α2 are the only human interferons that are naturally glycosylated. Naturally occurring human interferon-β is a glycoprotein with an apparent molecular weight of 25 kDa. The peptide structure has 166 amino acid residues and has a molecular weight of 17.5 kDa (corresponding to the nonglycosylated form). The predominant glycan structure of HuIFN-β consists of a biantennary complex-type structure attached to Asn-80.

Human interferon-β is secreted by fibroblasts in response to viral infection or to double-stranded RNA. The physiological mode of action of HulFN-β is not well understood. Interaction with a cell surface receptor causes the activation of a signal cascade that can affect the transcription of a number of specific genes. *In vitro* studies have shown the ability of HulFN-β to induce or suppress over 20 gene products. Its beneficial effect as a therapy for multiple sclerosis is ascribed to one of the following: (a) regulation of T'-cell function; (b) modulation of production of cytokines causing anti-inflammatory effects; (c) regulation of T-cell migration and infiltration.

Human interferon-β has antiviral and antiproliferative properties and has been used in chemotherapy in the treatment of certain types of solid tumours. However, results of this type of treatment have been disappointing. Pharmacokinetic evidence indicates that the problem might be the short half-life of HuIFN-β in the bloodstream, suggesting that poor performance in cancer trials may be caused by lack of sustained delivery to the tumour site. A solution to this limitation of delivery may be the use of gene *therapy* to the tumour site.

HuIFN-β was originally produced from large-scale cultures of human foreskin fibroblasts (such as FS-4) by a process termed 'superinduction' which involves induction with a synthetic polynucleotide (such as polyl:polyC) followed by a regime of antibiotic addition to maximize the stimulation of interferon synthesis. The rationale behind the superinduction process is to decrease the cellular con centration of an inducible protein repressor which normally causes the breakdown of interferon mRNA. The absence of the repressor can increase (×4) the half-life of IFN mRNA. The overall effect of this is to delay the syn thesis of IFN-β but to increase the total productivity 100-fold. The fibroblast cells are generally grown in microcarrier cultures to support growth to high cell densities. This is a procedure adopted by several companies with production occurring in large-scale batch cultures (up to 4000 liters).

In 1993 the use of recombinant HuIFN-β was approved in the treatment of multiple sclerosis (MS). This is a chronic inflammatory disease of the central nervous system that is autoimmune in origin. The result is a deterioration of nerve cells due to a T-lymphocyte-associated breakdown of the myelin sheath of the nerves. HuIFN-β has been shown in clinical trials to reduce the progression of MS in patients. This created renewed interest in the production of interferon. The isolation of the HulFN-β gene allowed the construction of genetically engineered *E. coli* and CHO cells for enhanced productivity of the protein. The specific productivity from these transfected cell lines is much greater than from the previously used superinduction process. There are currently three recombinant HuIFN-β preparations available for MS treatment—Betaseron (HuIFN-β1b), Avonex (HuIFN-β1a) and Rebif (HuIFN-β1a). Betaseron, which was approved in 1993 to treat relapsing, remitting MS, is a nonglycosylated form of recombinant HuIFN-β with a deletion of Met-1 and a Cys-17 to Ser mutation and produced in *E. coli*. The other two products are both glycosylated forms of HuIFN-β produced from transfected Chinese hamster ovary (CHO) cells in culture.

There are several recognized advantages of using the glycosylated HuIFN-β (type-1a) compared to the nonglycosylated form (type-1b). Firstly, it has higher biological activity. Avonex has been shown to have a ×10 higher specific activity than Betaseron in three separate functional assays based on antiviral, antiproliferatve and immuno-modulatory properties. This allows a lower patient dosage (30 μg weekly for Avonex against 250 μg every other day for Betaseron). Secondly, there is a significantly lower production of neutralizing antibodies against the glycosylated form of the molecule. Up to 42% of patients treated with Betaseron become positive for neutralizing activity after 3 months as opposed to only 6% of Avonex-treated patients after 12 months. Kagawa et al. (1988) analyzed the sugar chains of rHuIFN-β derived from three different mammalian cell lines. rHuIFN-β derived from other transfected cell lines (mouse epithelial, C127 and human adenocarcinoma, PC8) show greater differences in glycan structure compared to normal HuIFN-β and CHO-derived rHuIFN-β In particular, Galα1- 3Galβ1-4GlcNAc structures present from C127 and PC8 cells are absent in normal and CHO-derived HuIFN-β.

CHO cell-derived rHuIFN-β showed the greatest similarity to natural HuIFN-β. The proportion of biantennary complex structure was slightly lower (68%) than analyzed in natural HuIFN-β with a slightly higher proportion of triantennary forms. This compares with a corresponding value of 95% bianterutary analyzed by in CHO-derived HuIFN-β the close similarity in structure between the natural and recombinant forms is important to ensure against adverse immunogenicity following long-term clinical treatment.

IFN-γ

Synthesis of natural IFN-γ by I-lymphocytes can be stimulated by a range of antigens or mitogens. The most widely used inducer is staphylococcal enterotoxin A which can stimulate the extracellular release of IFN-γ by human peripheral blood lymphocytes obtained from donors. However, significantly higher productivity of IFN-γ has been obtained front recombinant mammalian cells transfected with the isolated gene. Transfected CHO cells have been used for the large-scale production of IFN-γ. Interferon secreted from a recombinant cell is a single subtype, whereas primary cells secrete a heterogeneous mixture of subtypes.

Plasminogen Activators

Thrombosis is one of the major causes of premature death in the Western world. This occurs by the deposition of fibrin in the circulatory system which results in a blockage of blood flow. Under normal physiological conditions the formation of fibrin is controlled by the clotting cascade which is activated during wound healing. The clotting cascade is in equilibrium with the process of fibrinolysis, which is a cascade of reactions culminating in the proteolytic cleavage of fibrin and which should prevent the undesirable formation of fibrin clots in the bloodstream. These include the 'kringle' domain characterized by looped structures held together by three disulfide bridges.

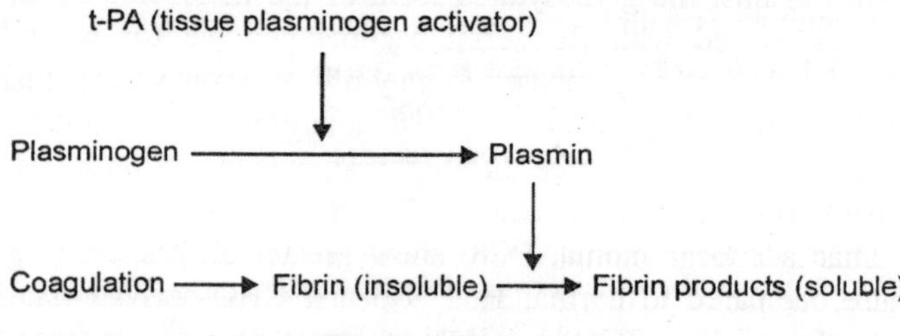

Fig. 13.4. Fibrinolysis.

The EGF domain is so named because of an amino acid sequence similar to one found in epidermal growth factor. The 'finger' domain is important in binding to fibrin whilst the serine protease domain contains the active site for proteolytic cleavage. The large-scale production process of tissue plasminogen activator (t-PA) was originally developed by Genentech and involves synthesis from transfected CHO-K1 cells. High specific rates of production were obtained from these cells by the incorporation of strong promoters for the transfected gene. In 1987, t-PA was one of the first recombinant products derived from mammalian cell culture to be approved for human therapy. For this regulatory approval it was necessary to show that each batch of cultures in the production process produced a consistent yield of t-PA and with the same degree of glycosylation. The final purity of t-PA from this process exceeds 99%.

The effectiveness of this t-PA has been compared with alternative compounds with plasminogen activator properties. These are streptokinase (produced from *Streptomyces*) and urokinase (extracted from urine). Clinical trials have shown that the higher specificity of t-PA for fibrin translates into positive benefits in the treatment of cardiovascular disease.

Blood-clotting Factors

Hemophilia is a sex-linked genetic disease which is characterized by an inactive clotting cascade in the blood. This results in an inability to form fibrin which is necessary for wound healing. The most common forms of clotting disorders of this type are hemophilia A and B which are caused by the absence of factors VIII and IX respectively

Therapeutic treatment has been performed routinely since the 1940s by regular administration of the appropriate factors to hemophiliacs. The clotting factors have been prepared as concentrates derived from the large-scale fractionation of pooled human blood plasma. Although this approach was successful for a number of years, the use of pooled blood sources is vulnerable to contamination with viruses such as hepatitis B and HIV. Although these viruses can be screened for and inactivated in human blood, the problems have encouraged the search for alternative sources of purified factors VIII and IX.

Factor VIII

This is a large complex glycoprotein (Mr = 265 kDa). In 1984 the gene for factor VIII was cloned, which was a significant achievement because of the size of the gene. The factor VIII gene is 186 kilobases long and consists of 26 exons and 25 introns. The cloning involved the ligation of a series of overlapping DNA fragments prepared from genomic and cDNA libraries, aligned to correspond to the full length of the gene, but without introns. Of particular significance in this strategy was the construction of a genomic DNA library from an abnormal human lymphoblastoid cell line with a karyotype containing four copies of the X-chromosome. Because the factor VIII gene is located on the X-chromosome this cell line is enriched with the required gene.

Transfection of a mammalian kidney cell line (BHK) with an expression vector containing the constructed gene allowed synthesis of the biologically active molecule with the correct tertiary folding and glycosylation. The recombinant factor VIII secreted by these cells can be stabilized by the addition of von Willebrand factor. The two factors are normally found as a combined protein complex in blood plasma and are both essential for blood clotting *in vivo*.

The large-scale process for factor VIII production presently operated by Bayer involves the use of a BHK cell line containing approximately 150 copies of the factor VIII gene per cell. The bioprocess involves continuous perfusion over 6 months using a cell retention system. Purification of factor VIII involves anion exchange, immunoaffinity and gel filtration combined with a viral inactivation step. The immunoaffinity chromatography is highly selective for factor VIII but subsequent steps of purification are necessary to ensure the complete removal of any monoclonal antibodies that could leach from the column.

Factor IX

Factor IX is a plasma glycoprotein (Mr = 57 kDa) secreted *in vivo* by hepatocytes. This is also known as Christmas factor, named after the first family diagnosed with the clotting deficiency. The gene for factor IX was originally cloned in 1982. Because the protein requires glycosylation and γ-carboxylation for full activity, a rat hepatoma cell line was used for expression of the gene. This cell line contains the enzymes for the post-translation modifications and was found to be an efficient one for the synthesis of recombinant factor IX.

Erythropoietin

The kidney is the source of synthesis of the glycoprotein hormone, erythropoietin (EPO), which is required to ensure continuous red blood cell production in the bone marrow (erythropoiesis). The absence of EPO in the bloodstream which may be caused by kidney failure will result in the impairment of red blood cell production and consequently, anemia. This condition is treatable by administration of exogenous EPO. Elevated levels of EPO may also be required after extensive blood loss.

EPO is a glycoprotein (Mr = 30-35 kDa) consisting of a 165 amino acid polypeptide chain, two disulfide bonds, three N-linked glycans at Asn-24, -38 and -83 as well as one O-linked glycan at Ser- 126. The unusually high proportion of carbohydrate (40% by weight) forms a significant component of the overall structure of the molecule as can be seen from the three-dimensional projections of the nonglycosylated and glycosylated molecules. The carbohydrate component is important for full activity *in vivo*. Partially glycosylated EPO is less effective clinically because it is removed rapidly from circulation by specific receptors present in the liver. CHO cells transfected with an expression vector containing the EPO gene have been produced by Amgen (USA). The resulting recombinant EPO is effective in the treatment of anemia associated with chronic kidney failure and was licensed for clinical use in 1988. This recombinant hormone has now been extremely successful in the treatment of patients with chronic kidney failure that results in anemia.

The bioactlvity of EPO is based upon the binding of a portion of the peptide structure with specific receptors on the surface of the red blood progenitor cells in the bone marrow. Although the glycan component does not affect this interaction directly, it is important to ensure adequate residence time in the bloodstream. Each of the N-glycans can be tetra-sialylated and the O-glycan may be di-sialylated giving a maximum of 14 sialic acid groups per EPO molecule. This results in a low pl (= 4.0). It has been shown that a reduction in the sialylatlon of EPO causes a proportional decrease in the serum half-life as a result of the removal of the glycoprotein from the circulation by the as aloglycoprotein receptor in the liver.

Egrie and Browne (2001) recently developed a novel form of EPO designated darbepoetin alfa or novel erythropoiesis-stimulating protein (NESP). This is a hyperglycosylated form of recombinant EPO in which two new N-glycan sites were introduced into the molecule by site-directed mutagenesis at amino acid positions 80 and 88. Each of the additional N-glycans increases the molecular weight of the glycoprotein by approximately 3.3 kDa and may contribute up to four additional sialic acids. Thus, the molecular weight of NESP is 37.1 kDa with a maximum of 22 sialic acid groups and a pl of 3.3.

Through clinical trials it was shown that the serum half-life of NESP is 26.3 h, which is about 3-fold greater than rEPO.

- rEPO (three N-glycans);
- an analog with four N-glycans;
- NESP.

Even though the binding capacity of EPO to its receptor decreases with sialylation, the net effect is to enhance the biological activity *in vivo*. This is shown by the gradual increase in the red blood cell count (hematocrit) of three groups of mice, each treated for 30 days with one of the EPO isoforms containing a specific number of sialic acids. The most effective therapeutic isoform of EPO has the highest number of sialic acids. The value of NESP as a therapeutic agent is that the frequency of administration to patients can be decreased

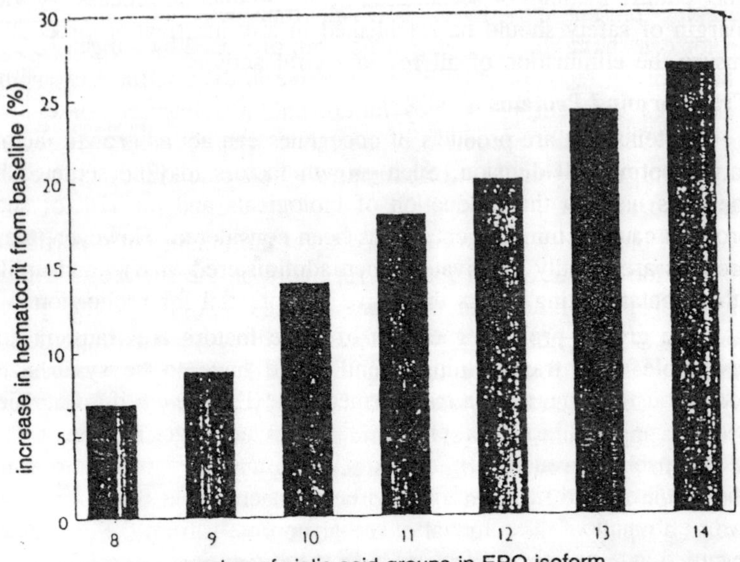

Fig. 13.5. The biological activity of each isoform of EPO after a 30-day treatment.

significantly for the same clinical effect. However, there is still some uncertainty over the long-term immunogenic response to NESP, which is possible because of the altered amino acid sequence.

Risks Associated with Cell Culture Products

There are important safety considerations and risk factors that must be considered before a biological product is accepted for clinical use. There are four main identifiable risk factors associated with animal cell-derived products. The risks are associated with viruses, transforming proteins, DNA and the characteristics of the producer cell line.

Viruses

The presence of endogenous viruses associated with a cell line poses a recognizable risk for product contamination. Retroviruses are of particular concern because of their tumourigenic potential. For many processes, a variety of chemical and physical treatments during product isolation may be sufficient to inactivate any contaminating viruses. However, in order to be certain, it is essential to validate the efficiency of the purification process by following the loss of viability of a range of known viruses added at various stages of isolation. Viral assays may be performed to detect the presence of contamination in the cell line, culture medium or serum used in the production process. A wide margin of safety should be established in any inactivation process to ensure the elimination of all residual viral activity.

Transforming Proteins

Proteins that are products of oncogenes can act as growth factors in promoting cell division. Such growth factors may be secreted by the cells used in the production of biologicals and the risk of such proteins causing tumourigenesis has been considered. However, these proteins are rapidly inactivated when administered *in vivo* and unlike other contaminating agents they have no potential for replication.

The growth-promoting effects of these factors are transient and reversible. The transforming agent would have to be synthesized continuously to maintain a transformed cell. Thus, even if co-purified with the required biological product, such contaminating proteins would be incapable of causing or maintaining the transformation of cells *in vivo*. Consequently no specific tests are considered necessary to ensure their elimination. The high purity of the final product is generally regarded as a sufficient safeguard against the risk of such contaminating proteins.

Residual Cellular DNA

The DNA content of animal cell-derived therapeutic products causes the greatest concern because of the potential for carrying fragments of oncogenes. The oncogenes may have the potential for replication and could lead to tumourigenesis. These genes are expressed in cancer cells and ensure their continuous growth potential. In order to eliminate the risk of product contamination with oncogenes, the purification process should reduce the residual DNA content to a minimal and safe level.

Current methods of analysis allow the determination of residual DNA in. an isolated cell culture product to a concentration of at least 1 pg/ml. A DNA content of $<$ 10 pg/dose of an injectable product is sufficiently low to assess the risk of tumourigenesis as negligible. This level of DNA is several orders of magnitude lower than the minimum required for cell transformation *in vitro*.

Safety of the Cell Line

Although continuous cell lines have good growth characteristics in large-scale culture, they were restricted from use in licensed production processes for many years because of the perceived risk of transmission of tumourigenic agents. These cell lines are known to have activated oncogenes and many harbor endogenous viruses. However, the extensive characterization of certain cell lines and the ability to reduce the detection limits of known contaminants such as DNA has enabled a relaxation of the blanket restriction against the use of these cells. Greater emphasis has now been placed on the ability to prove that the final products are not contaminated and are therefore risk-free.

In the case of production of recombinant proteins it is essential to use continuous cell lines. Protocols for genetic engineering require continuous cell culture for over 100 population doublings before a suitable highproducing cell line is obtained. Normal diploid cell lines would enter the senescence state well before such procedures were complete. Many recombinant products from continuous cell lines have been accepted for clinical use and approval is related to the maintenance of stringent procedures of culture and product purification.

Tissue Engineering

The ability to re-constitute human tissue from combinations of cell types grown in culture is an important prospect for future therapeutic treatment associated with organ failure. Up to now the most successful application of tissue engineering is in the re-constitution of skin following severe burns.

Artificial Skin

The reconstitution of human tissue is important for skin grafting. Artificial skin can be formed from two layers derived from cultured human cells:

- a dermal-equivalent formed from fibroblasts;
- an epidermal-equivalent which is layered on the dermal surface.

To construct a dermal-equivalent, fibroblasts are dissociated from a tissue biopsy and mixed with predetermined proportions of collagen in culture medium. The mixture is poured into a mold, the shape of which determines the geometry of the constructed tissue. The fibroblasts cause the condensation of the collagen into fibrils to form a gel-like matrix which excludes most of the surrounding fluid. During a 6-day culture period the cells attach to the collagen forming a contracted matrix of tissue-like consistency.

An epidermal-equivalent developed from keratinocytes is then layered on the surface of the dermal-equivalent. The epidermal cells attach to the collagen substratum to form a continuous sheet in which cells can differentiate and form multilayers. After 10 days the epidermal cells may be exposed to air to allow the formation of an outer protective layer (stratum corneum) which consists of a layer of dead cells. The bi-layer of cells increases in thickness over time and results in a controlled deposition of matrix proteins. This living bi-layered skin substitute is created to heal wounds and has been used successfully for grafting to burns patients. The process has been demonstrated to be safe and effective for therapeutic treatment. Clinical trials of one such product has been effective in the treatment of acute and chronic wounds without immunosuppressive therapy.

The keratinocytes and fibroblasts are derived from neonatal foreskin tissue and lack antigen presentation that would cause an adverse immune response. The skin substitute can be cryopreserved under liquid nitrogen and reconstituted quickly when required.

Artificial Organs

Attempts to construct organs other than skin equivalents have met greater technical difficulties. Multiple cell types require more complex scaffolds and an extracellular matrix to support the functional relationship between cells. A further level of complexity is that multiple cell layers require a nutrient supply equivalent to blood capillaries *in vivo*.

However, the use of cultured cells to replace damaged cells *in vivo* offers some immediate and promising possibilities. This concept has been developed for pancreatic cells capable of secreting insulin to replace damaged islet cells in patients with diabetes. One problem is how to protect the transplanted cells from immunogenic rejection. A possible solution is to encapsulate the cells in a matrix such as alginate or agarose. This would protect the cells from immunological rejection but still allow the free flow of nutrients and small secreted cell products. Thus, hormone replacement therapy may be possible by development of such artificial organs.

Cells as Products

This section describes the various ways that cells can be useful in procedures for toxicity testing and for various clinical or therapeutic applications.

Drug Screening and Toxicity Tests

The use of whole animal tests to establish the potential toxicity of compounds has been routine for many years. However, there are a number of disadvantages and objections to the continuation of these tests. The cost and variability of results is considerable. Also, there are growing moral objections to the use of animals for such purposes. Mammalian cell cultures can offer the basis for alternative test systems that may overcome some of these disadvantages. Cell culture. tests are rapid, allow more efficient screening of novel compounds and sometimes can allow the identification of metabolic targets of inhibition. These test systems can be designed to evaluate various effects:

- reduced growth rate;
- breakdown of membrane permeability;
- tissue specificity of response;
- ability to metabolize toxic compounds;
- simulated wound healing;
- damage repair by use of artificially constructed tissue;
- genetic effects/mutagenicity, i.e. interaction with DNA.

Cell Therapy

The ability to expand and differentiate hematopoietic stem cells *ex vivo* has many clinical applications. There is a high turnover of hematopoietic cells *in vivo* as a result of continuous growth and replacement. This makes the cells vulnerable to destruction by any cytotoxic drugs used in chemotherapy to eradicate residual tumour cells. Removal of bone marrow cells prior to chemotherapy allows isolation and expansion of the pluripotent stem cells to provide a source of mature hematopoietic cells following chemotherapy.

The *ex vivo* cell expansion can be manipulated by careful culture control of stem cells to allow the production of a cell population enriched with specific cells lines. This procedure can be useful for the production of specific T-lymphocytes or natural killer cells that may be required for cell-mediated immunotherapy. The hematopoietic stem cells can also be an appropriate target for somatic gene therapy as described in the next section.

It is often difficult to identify the stem cells from a mixed population, although one method that has been widely adopted is immunopurification with anti-CD34 antibodies. The cell population isolated with a cell surface CD34 antigen is enriched with hematopoietic stem cells. These can be expanded in a medium containing a controlled balance of cytokines. It has been found that various cell culture parameters including oxygen, pH, osmolarity and cytokine composition can influence the extent and direction of differentiation.

Gene Therapy

The development of methods for transfecting exogenous genes into mammalian cells has been extremely important for producing cells capable of expressing large quantities of therapeutic proteins in culture. However, the transfected cells themselves may also be used directly as a treatment for certain genetic diseases associated with inborn errors of metabolism.

The concept of gene therapy is that a missing or faulty gene is replaced by a normal, working gene. The process involves the transfection of a specific gene into cells of a patient with an identified and well-characterized genetic disease. The gene can be introduced into cells inside the patient (*in vivo*) or outside the patient (*ex vivo*). For the *ex vivo* process the cells that are removed are then re-introduced into the patient when they express the normal protein which alleviates the metabolic defect.

An example of the successful application of this is in the treatment of severe combined immunodeficiency (SCID) associated with a defective copy of a gene on the X-chromosome required for the expression of the enzyme adenosine deaminase (ADA). Treatment by gene therapy involves the isolation of bone marrow stem cells from the patient followed by *in vitro* infection of the cells with a retrovirus constructed to carry the ADA gene. The transduced stem cells are then re-introduced into the bone marrow of the patient where they can proliferate and differentiate into immunocompetent cells. This treatment has been carried out on a small number of patients.

Treatment of this type may be applied to other diseases associated with bloodborne cells. For example, the therapeutic procedure is suitable for treatment of disorders associated with hemoglobin, such as sickle cell anemia of thalassemia. Sickle cell anemia arises from a single point mutation in the β-globin gene. The thalassemias represent a collection of glohin abnormalities, arising from a reduction in α-chain synthesis (α-thalassemia) or in β-chain synthesis (β-thalassemia). The hematopoietic cells can be isolated from the bone marrow and transfected with normal globin genes using retroviral vectors. These cells when re-introduced into the bone marrow will provide a continuous source of normal hemoglobin.

Other potential disease targets for gene therapy include Lesch-Nyhan syndrome, which is an X-linked recessive deficiency of the hypoxanthineguanine phosphoribosyltransferase (HGPRT) enzyme and phenylketonuria, which is associated with deficient phenylalanine hydroxylase enzyme. There is some debate as to the most effective and safe vector to be used in gene therapy. There art, two perceived dangers:

1. There may be a severe immune reaction to the applied viral vector;
2. The viral vector may integrate into the DNA of a cell in a way to cause the activation of a cellular proto-oncogene.

The four viral vectors that have been considered for gene therapy are:

- Retroviruses;
- Adenoviruses;
- Adeno-associated viruses;
- Herpes simplex virus.

These viruses vary in their characteristics and requirements. The ideal viral vector shows efficient entry into cells followed by high gene expression for long periods. None of these vector types is ideal in all situations.

Conclusion

The number of novel recombinant proteins to have received regulatory approval as human therapeutic agents has risen sharply in the last few years. As a result of opportunities arising from the human genome project, an understanding of the role of specific genes has given rise to a large number of potential mammalian proteins to be used as therapeutic agents. At the present time more than 200 drugs

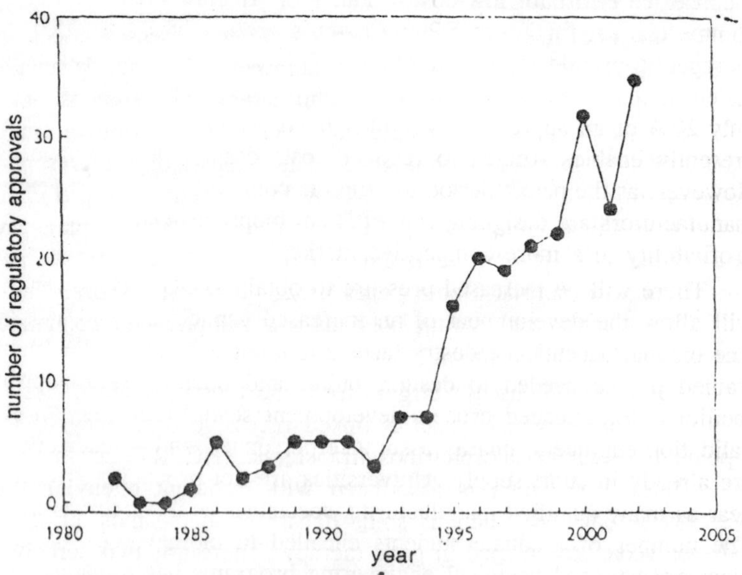

Fig. 13.6. New biotechnology drug and vaccine approvals per year.

derived from mammalian cell cultures are in clinical trials, and about 70 of those drugs are expected to come to market by 2006. A high proportion of these are humanized monoclonal antibodies and many are targeted as anticancer agents.

The need to produce this large number of glycoproteins under GMP (Good Manufacturing Practice) conditions will result in enormous pressure on the available manufacturing facilities. There is a well-recognized shortage of pharmaceutical and contract manufacturing facilities for the large-scale production of the recombinant protein products targeted as potential therapeutic agents. The current worldwide cell culture capacity, estimated at 450,000 liters, is almost fully utilized. To meet this demand biopharmaceutical companies and contract manufactures are now attempting to triple their production capacity. However, new bioprocessing plants take 3-5 years to design, build, and certify. This has already resulted in shortfalls in production. For example, it is reported that the lack of manufacturing capacity for the biopharmaceutical company, Immunex, resulted in a loss of revenue of more than $200 million for its arthritis drug in 2001.

The cost efficiency of a production process becomes more important as products move out of the patented domain of specific companies. Competition amongst generic recombinant protein producers will force a concerted effort for low-cost production. An example of this is the commercial production of EPO. Amgen's production process for this therapeutic protein is an inefficient multiple-culture system based on roller bottles. The lack of process control leads to the recovery of only 20% of an appropriately glycosylated product. Patent protection presently enables Amgen to recover costs despite this inefficiency. However, as the patent period for EPO is coming to an end so generic manufacturers are designing cost-efficient hioprocesses that will allow profitability in a more competitive market.

There will be increased pressure to obtain skilled manpower that will allow the development of an increased capacity for production. The biopharmaceutical industry faces a looming shortage of the highly trained people needed to design, build, and operate manufacturing facilities. Experienced process-development scientists and engineers, validation engineers, quality assurance personnel, and plant managers are already in short supply. Universities are not producing anywhere near as many qualified manufacturing specialists as the industry needs. The number of graduate students enrolled in biology, cell biology, biochemistry, and chemical engineering programs has remained flat, whereas demand has increased with the addition of new capacity.

14

PRODUCTIVITY

Many healthcare products can be derived from fractionation of blood or extraction from human tissue. However, it is difficult to ensure that such products are free from contamination, particularly by viruses. This risk can be reduced considerably by using well-characterized animal cell lines that are shown to be free of contaminating viruses. As an example, factor VIII used for the treatment of hemophilia was purified from pooled human plasma in the early 1980s. This process was discontinued when samples were found to be contaminated with human immunodeficiency virus (HIV). The safety of the present recombinant product is now ensured by careful screening of the mammalian cell line used in the large-scale bioprocess.

Genetically engineered mammalian cells can overexpress and secrete glycoprotein products into the culture medium. Protein extraction from the culture medium is relatively efficient particularly if a low-protein culture medium is used. In comparison, secretion of protein from genetically engineered bacteria can be more problematic. The synthesized protein usually accumulates as insoluble aggregates within bacterial cells and so requires extraction from the cell lysate. This is more difficult because of the high content of other proteins (including some endotoxins) that must be completely removed from the final product. Thus, the recovery process from the cell lysate is more complex and the final yield of purified product may be lower.

HOW TO PRODUCE BIOLOGICALS FROM CELL CULTURE

The first obvious questions to ask before setting up a cell culture process for production are:

(i) What product do I want?
(ii) How pure do I want it?
(iii) How much do I want?

The type of process, strategy and design will depend upon the answers to these three critical, questions.

What product?

The type of product required is the key to deciding which type of cell to use for production. For simple proteins such as insulin or growth hormone that have low molecular weights and no attached carbohydrate groups, a genetically engineered *Escherichia coli* culture producing the recombinant protein may be preferred. Bacteria grow faster and under more robust conditions than animal cells. However, there are limitations of using bacteria for the expression and production of more complex mammalian proteins. The most serious consideration is that there are a number of post-translational processes that are necessary to modify proteins before they become biologically active:

- proteolytic cleavage;
- subunit association;
- chemical derivatization;
- glycosylation;
- phosphorylation;
- fatty acylation.

In particular, glycosylation has received a lot of attention. Glycosylation is the intracellular process that involves the addition of a carbohydrate group on to the synthesized protein. The process occurs in the Golgi apparatus prior to secretion through the cell membrane. The carbohydrate component of glycoproteins is often essential for biological activity. Proteins that are under-glycosylated are often cleared from the bloodstream too quickly.

Prokaryotes (such as *E. coli*) are not capable of glycosylation. Yeast cells may be genetically engineered to produce a protein from a transfected mammalian gene. Yeast cells can glycosylate proteins but the carbohydrate added may be different from that in the authentic mammalian glycoprotein and this could influence the therapeutic activity of the molecule.

The carbohydrate content of naturally occurring glycoproteins can vary from 3% (e.g. immunoglobulin) to 40% (e.g. erythropoietin) of the total weight and can be important for many biological properties. Glycoprotein extracted from a culture may be in various glycoforms.

These are molecules having an identical protein structure but with differing carbohydrate additives. This type of variation may be unacceptable for some applications, such as for therapeutic use, where a single glycoform may be required. The extent of glycosylation of proteins produced in culture depends on the conditions. The optimal conditions for full glycosylation are often unknown but it has been shown that reduced glycosylation may occur at the end of the growth phase and during the stationary phase of culture.

In most cases, a genetically modified cell line will be required to obtain a specific glycoprotein at a reasonable yield, using the principles. The most commonly used host cell lines for such production are Chinese hamster ovary (CHO) and baby hamster kidney (BHK) because their genetics and growth characteristics have been well characterized. However, in some cases a mammalian cell line may be found that will produce the required glycoprotein product without any genetic manipulation. An example of this is interferon (alpha or beta) production from human fibroblasts. An induction process has been developed which allows high productivity from these cells. The production of animal viruses from cell culture requires some consideration of an acceptable host cell. In this process it is important to know which cell line is susceptible to the particular strain of virus required. Also, the mode of culture may be important. For example, the extent of viral infection is often greater in anchorage-dependent cells attached to a substratum.

How pure?

If a product is required for human injection—as a therapeutic agent or as a prophylactic viral vaccine—then high standards of purification are needed with an absolute requirement to ensure the absence of known *risk factors*. If a product is required as a diagnostic agent or for laboratory use, then the conditions of purification may be less stringent. Electrophoresis is a powerful method for discriminating between proteins. Thus the visualization of a single stained protein band by gel electrophoresis is a good indicator of the purity of a protein.

The purity requirements for the final product may influence not only the purification process (downstream processing) chosen but also the conditions of culture. For example, purification is easier if the cell product is secreted into a low-protein or protein-free culture media. Also, it may be necessary to avoid antifoam or shear protective agents as these can foul membrane-based purification processes. The timing of product harvest may be important because of the risk of product exposure to proteases or glycosidases, that may arise from cell lysis.

These enzymes will cause breakdown of the protein backbone or carbohydrate group of a glycoprotein product.

How much?

Deciding the quantity of cell product required over a set period of time is vital to planning a production process. This is equally important for a laboratory preparation as well as a large-scale commercial enterprise. The importance of this question to developing a process strategy can be best illustrated by an example of the production of a hypothetical glycoprotein.

The biotechnology company, Candogene, planned to make 10 kg of Biolikin from a cell with a product yield of 1 pg/ cell-day. Could this requirement for 10^{16} cell-days of production be met by their stirred tank reactor at 10^6 cells/ml? The strategic planners at Candogene determined that this would require a day's run on a 10 million (10^7)-liter fermenter and abandoned the project as beyond their scope. However, an alternative company decided to take up this project by a three-stage plan.

1. They improved the specific productivity of the cell line ($\times$ 100) by gene amplification. This reduced the required culture capacity to 10^5 liters.
2. They decided to use a perfusion culture that increased the final cell concentration to 4×10^7/ml. This reduced the required culture capacity to 2500 liters.
3. They decided that there was no need to make the product in 1 day but could continue the process over 100 days. This reduced the required culture capacity to 25 liters, which was well within the fermenter size available in their small laboratory.

This example illustrates that missed opportunities can arise from poor or no planning and by not asking the right questions from the onset of the work.

Scale of Production

Animal cell products generally have a high value but a low demand in terms of quantity. This can be illustrated by figures available for erythropoietin, which is one of the successful recombinant products used as a clinical agent. According to a figure from 1998, the total US demand for this therapeutic agent was satisfied by 1 kilogram per year and the product was sold for \$1600/mg. This demand can be accommodated by a 10,000 liter bioreactor with a production level of 500 units/ml after a 10 day period of cell growth. Typically, the specific activity of EPO is 130,000 units/mg. This principle applies to most animal cell bioprocesses, where the scale of homogeneous suspension cultures is rarely above 10,000 liters and for nonhomogeneous microcarrier cultures rarely above 1000 liters. The three major technical problems of scale up of unit processes are mass transfer, mixing times and shear forces.

Mass transfer

Oxygen must be provided to meet the metabolic demand of the cells without high stirring rates or intense gas sparging that may cause cell damage. The damaging effect of gas sparging may be reduced by additives such as Pluronic.

Mixing

If the cultures are fed with nutrients then the mixing rate should be adequate to prevent concentration gradients that could damage cells or cause uneven growth.

Shear forces

If the stirring rates are not too high then the cells will not be damaged by shear forces. However, the stirring rate must be sufficient to allow efficient mixing and mass transfer.

Type of Bioprocess

Most large-scale bioprocesses presently in operation are designed as fedbatch cultures. The advantage is simplicity and ease of scale-up. High cell densities may be obtained if an appropriate mixture of components is fed to the cells at strategic intervals.

The alternative mode of culture that is used in around 10% of bioprocesses involving animal cells is perfusion. This has the advantage of allowing high cell densities by a continuous supply of nutrients and removal of cell-free products from the bioreactor. This removes the products from long contact with the cells that may release degradative enzymes such as proteinases and sialidases. There is a high productivity from such a process and it can be maintained for long periods of time. The disadvantage of the perfusion bioprocess is the high initial starting cost. Because of this, the perfusion mode is best suited for the production of proteins that are required in relatively large quantities ($>$100 kg/year).

Alternative Methods for Multiple Products Required in Small Quantity

Monoclonal antibodies are a special category of products because they are required in relatively small amounts but in multiple types with differing antigen specificities. The standard method of production has been injection of hybridomas into the peritoneal cavity of mice. The growth of the hybridoma releases monoclonal antibody into the ascites fluid which can be extracted from the mouse.

The production of monoclonal antibodies by mice has come under increasing criticism because of the ethical issues posed by the use of laboratory animals and particularly given the availability of alternative methods. In most European countries regulatory approval has been withdrawn except, when it can be shown that the use of animals is the only means of production. Additionally the ascites method presents problems for product purification. The overall protein content of ascites fluid is high, posing difficulties in obtaining pure monoclonal antibodies. Furthermore, the ascites fluid contains antibodies secreted by the host mouse, which are impossible to separate from the desired monoclonal antibody. Thus the final product following purification stay have residual activity that will interfere with the activity of the monoclonal antibody.

In vitro production of monoclonal antibodies can occur by small hollowfiber bioreactors which can be run continuously for several months. The antibody productivity from such a bioreactor run for 2 months has been shown to be equivalent to the use of around 20 mice. Depending upon the specific productivity of the hybridoma, this would typically produce 100-1000 mg of antibody. A cost analysis shows that these processes are broadly equivalent with a production cost of $1-2 per mg antibody.

The hollow-fiber bioreactor is a convenient system for continuous production of small quantities of antibody, although some problems can arise because of the high cell densities that occur. Limited gas diffusion may produce an oxygen deficiency which has been shown to reduce full glycosylation of the secreted immunoglobulin. An alternative is to use a simple small batch culture such as the gas-permeable bags that have been designed from 0.5-2 liters.

How to Purify the Final Product

The process of purification—often called 'downstream processing' —is dependent on the product and the degree of purification required. However, some general principles can be provided here.

Primary separation involves the removal of cells and debris from the medium. In the case of immobilized cell cultures, this can occur by sedimentation of the cell-loaded particles. For suspension cultures, cells are normally removed by low-speed centrifugation or by filtration. It is important in this primary separation to minimize cell lysis and ensure the removal of all cell debris as this can give rise to degradative enzymes which are capable of causing product breakdown. Chilling the medium (4°C) can also minimize the activity of these enzymes. Product concentrations arising from a cell culture are generally low—10-100 mg/l. So, the next stage of purification often involves increasing the concentration by a de-watering process. This can be accomplished by precipitation with a suitable agent such as ethanol, polyethylene glycol or ammonium sulfate. These precipitating agents can be specific for certain protein types and it is important to ensure that the conditions and concentration are appropriate for precipitating the protein required.

An alternative method of concentration is by ultrafiltration using a hollow-fiber cartridge, which can concentrate by 10-100-fold. The hollow-fiber cartridge should have a high-molecular-weight cut-off, so that the product, but not the cells, passes through the fiber. For example, with a 10-kDa cut-off, the concentrated 'retentate' will include all substances above that molecular weight while all lower-molecular-weight substances will be drained off as 'permeate'. To maximize the yield of the final product extracted from culture media, it is important to have as few purification steps as possible. This can be achieved particularly if highly selective fractionation procedures are available.

Bioselective adsorption methods are particularly valuable because of the selectivity for a single compound or group of compounds. For example, Protein A and Protein G have specificity for binding immuno-

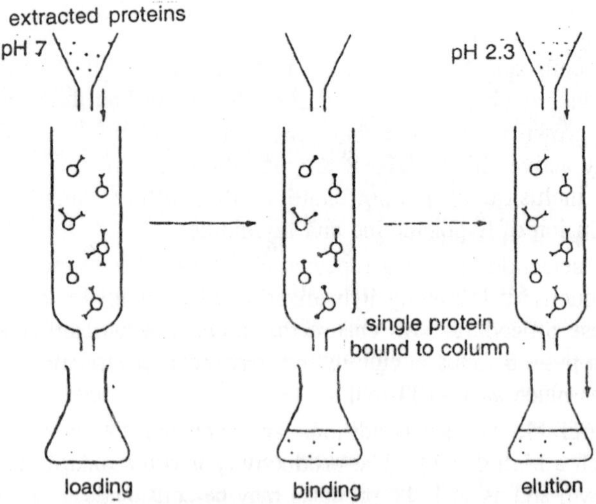

Fig. 14.1. Immunoaffinity purification.

globulins. Also, monoclonal antibodies are useful because of their affinity for particular compounds. The adsorption reagents can be prepared by immobilization of an antibody to a solid matrix such as Sepharose. This is suitable for column chromatography because of the open-pore structure which allows a reasonably high liquid flow rate. Reactive forms such as cyanogen bromide-activated Sepharose are available commercially and can be bound covalently to a protein ligand (such as an antibody). A protein extract is loaded on to the column at neutral pH, allowing a single protein (the antigen) to bind to the antibody sites. After washing the column with further neutral pH buffer, the antigen is eluted in a pure state with a low pH buffer.

Efficiency and Productivity of a Culture System

The production of biologicals from animal cells is akin to chemical synthesis in which the cells can be considered as catalysts promoting the conversion of relatively simple nutrient molecules into specific complex macromolecules. All genetic or biochemical manipulations of the cells are designed to convert them into efficient chemical catalysts —a process which may change the cells significantly from their original state.

Cell products are normally secreted into the culture media by viable cells. The final yield of the product will depend on:

(i) the specific productivity of each viable cell—expressed as μg of product formed per 10^6 cell-day;

(ii) the viable cell density of the culture ($\times$ 10^6 cells/ml).

Productivity can be maximized by ensuring the following.

1. The cells selected following transfection have the highest level of product synthesis. In some cases selective pressure while the cells are grown may ensure that genes are continuously amplified and fully active. In the DHFR amplification system, this may require the inclusion of methotrexate in the culture medium and the exclusion of hypoxanthine and thymidine.
2. Cell selection involving repeated cloning is also important for high productivity following hybridization of antibody-producing cells. These procedures will ensure that a cell line will be obtained for high protein yields in culture and may make productivity differences of as much as 10-100-fold.
3. Favourable culture conditions are required to ensure that cells reach a high density. The productivity is often independent of cell growth and as such the products may be considered to be secondary

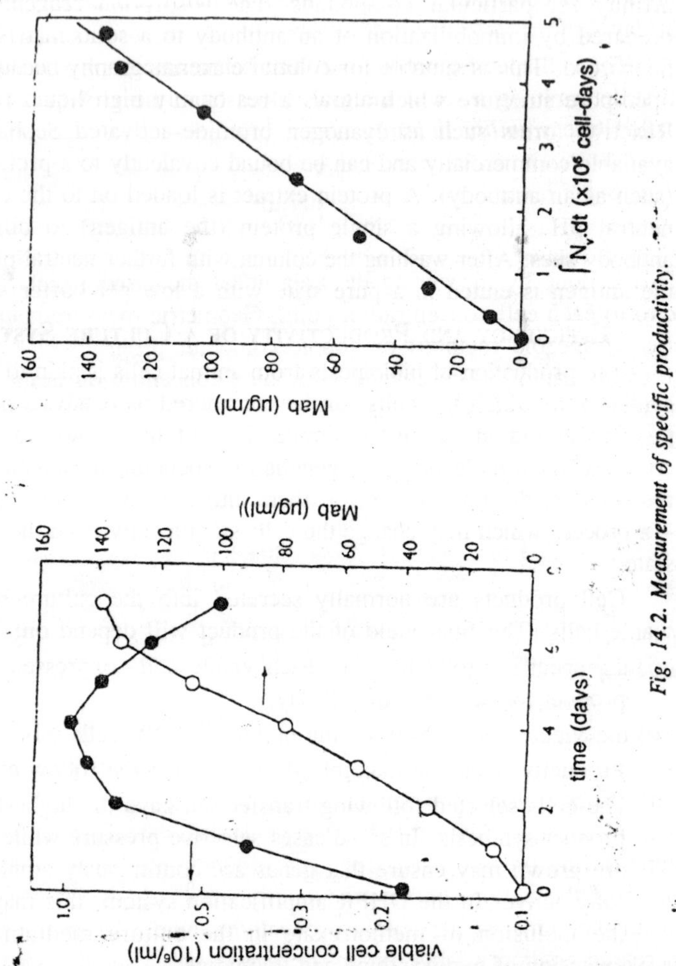

Fig. 14.2. Measurement of specific productivity

metabolites. This means that productivity will be directly related to the number of viable cells in culture and it is an advantage to maintain the cells in a viable condition for as long as possible through the growth and stationary phases. This may not be possible for some cells. If cell death occurs by apoptosis then the culture phase will shift quickly from a growth to a decline phase—a phenomenon typical of antibody-secreting hybridomas.

The monoclonal antibody (Mab) concentration increases to 140 μg/ml over the 7-day culture period. Notice that in this example production continues even after the maximum cell density is reached at day 4.

The cell-specific productivity can be measured from these data by first determining a set of integral values from the growth curve (often called 'the *viability index*'):

$$\int_0^\tau N_V dt$$

This is done by measuring the area under the growth plot from time zero to each cell concentration point. Geometric measurements can be used (trapezoid method) or simple computer programs are available (e.g. Sigmaplot). The change in product concentration is then plotted against the viability index and the specific rate of productivity determined from the slope.

$$Q_s = \Delta P \cdot \frac{1}{\int_0^\tau N_V dt}$$

This calculation is similar to that used for substrate uptake and by-product formation.

The gradient of the resulting straight line is the specific productivity (pg/cell-day or $\mu g/10^6$ cell-day). For the generation of some products such as monoclonal antibodies the specific production rate has been found to be constant throughout the culture of hybridomas. If production is growth related then this will be apparent from the plot by a steep slope during the growth phase. The production is independent of the phase of the culture and a productivity of 30 $\mu g/10^6$ cell-day can be calculated. Growth-dependent productivity would be indicated by an increased gradient of this plot in the growth phase and then a decreased gradient as cell concentration declines.

The productivity of a cell culture process may be expressed in several ways to give an indication of efficiency. Final cell densities can vary between typical values of 10^6 cells/ml for batch cultures, to 10^7 cells/ml for perfusion, to 10^8 cells/ml for hollow-fiber systems. However, you need to assess such comparisons with care. A higher final cell density in a perfusion or hollowfiber system may be obtained following the expenditure of a considerable volume of medium. High cell densities may occur only in part of the bioreactor system such as the extracapillary space of hollow fibers or the interior of cell-entrapping beads. The cell yield per total volume of expended medium may be a more valuable comparison to make between such systems.

However, even this can be misleading as an indicator of the overall process efficiency and it is important to consider all the factors involved in the production process. For example, a product released into a low-protein medium may eventually be extracted more efficiently than one released into a high-serum medium. This can have an important bearing on the economics of the production process. Thus, the ultimate indicator of the efficiency of an overall process is the total operation cost per unit weight of the purified product.

Cost of the Cell Culture Process

A cost analysis of a commercial operation involving animal cells will take into account various factors including labour, materials and capital cost. Generally the unit cost of production decreases with scale-up. This is usually an indication of a significant decrease in the labor cost as a percentage of the total unit cost. This is because the demands of managing a 1000-liter fermenter are less than 10 times that of a 100-liter fermenter—a fact that shows the value of a unit process as compared with multiple processes where the number of manipulations increases proportionally to the number of cultures in operation. Here the advantage of scale-up is shown by a dramatic decrease in the unit cost of the product with a × 100 increase in scale. Thus, 1 mg of Mab produced in a 10 000-liter bioreactor would cost 90% less than the equivalent cost in a 100-liter bioreactor.

At the smaller scale the labour cost is the major component of the total production cost. However, this decreases dramatically with increase is the scale of operation, from 80 to 10% of the unit cost in the example. On the other hand, material costs increase substantially as a proportion of the total cost during scale-up. This suggests that any saving in the cost of materials, such as the growth medium, could have a significant effect in lowering the unit cost of a product of a large-scale process. In order to reduce media costs, several approaches may be considered.

Omission of antibiotics, use of a lowered serum content or use of an alternative to fetal calf serum would permit significant cost reductions. Fetal calf serum can account for as much as 90% of the total medium cost. Serum-free media formulations may be suitable for the growth of specific cell lines although these formulations may contain certain hormones and growth factors, the combined cost of which may

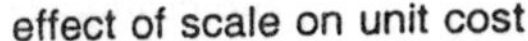

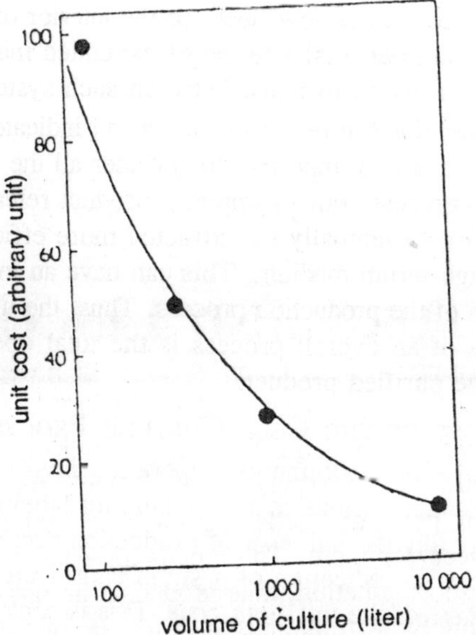

Fig. 14.3. Cost analysis of bioprocess scale-up.

exceed that of the serum supplement. However, a serum-free formulation may still be justified in a large-scale operation if there are benefits in product purification.

The optimization of the cell culture strategy can increase productivity and can offer an alternative to increasing culture volume. The choice of media perfusion rather than batch culture can lead to significant increases in cell yields. Continuous cultures generally have a higher productivity than batch cultures. Although perfusion may require the expenditure of larger volumes of medium, it may be possible to use diluted media or carefully controlled dilution rates to improve overall productivity. The formulation of optimal growth media or the development of a regime of selected nutrient addition and inhibitor removal may lead to routinely higher cell yields. Considerations of this sort in planning culture operations may significantly alter the economics of a process and reduce the overall unit cost of the final product.

15

Culture and Oncogenes

During animal development, cell division and cell differentiation are carefully regulated, but occasionally the controls that regulate cell multiplication break down and cells begin to grow and divide although the body has no need for them. These cells have the ability to divide without responding to regulation, thus producing groups of cells which can expand indefinitely. A tumour is such a mass of cells which have escaped from the normal mechanisms controlling cell growth. Tumours are also called *neoplasms*, which means literally 'new growth', but has come to mean abnormal new growth of tissue. Likewise tumour cells are sometimes called *neoplastic cells*, or *hyperplastic cells*. The term *cancer* is used to describe a large group of diseases which arise because of changes in growth and behaviour of cells.

The study of neoplastic cells and the cancers they produce is of fundamental importance. Cancers are amongst the most feared medical conditions and studies on how they arise may enable the development of strategies for preventing and curing these conditions. As we will learn later, the study of neoplasia has, to a large extent, depended upon using cells cultivated *in vitro*. This alone justifies inclusion of a chapter on cancer in this text. There is however a further justification. Neoplastic cells are cells that have escaped from normal regulatory control. The study of these cells may, therefore, provide insights into how cell growth and division are regulated in normal cells. This has potential practical implications in the use of cultivated animal cells to make biotechnological products such as specific mammalian proteins. The enhanced growth rates and ability to grow in relatively simple media, both features of neoplastic cells, are, of course, desirable features for the use of animal cells cultured *in vitro* for commercial purposes.

Cancer and Cell Transformation

The unpleasant symptoms of cancer result from the unlimited growth of the neoplastic cells, which results in an increase in the bulk of tissue, impinging on the structure and function of adjacent tissue. In the later stages of cancer, the cells of the tumour can penetrate and invade adjacent tissues, spreading from the primary site to distant sites in the body. It is this ability to spread (metastasize) that characterizes malignant tumours. As a cancer progresses, there are changes in the histology of the tumour which include the tendency of the cells of the tumour to lose their differentiated properties. From a prognostic stand-point, patients with poorly differentiated tumours generally have a lower survival rate than those with well differentiated tumours.

Despite tremendous efforts, there has been little improvement in the overall survival of patients with cancer over the last 30 years. Whilst changes in the management of lymphomas and leukaemias have produced a dramatic increase in survival, these contribute little to the overall picture. Around one person in three will develop cancer at some time during his or her life. It is likely that the overall incidence of cancer will continue to rise as the average age of the population increases. Investigations into the causes, prevention and treatment of cancer is, therefore, a major component of scientific research.

There are many ways of studying cancer ranging from clinical trials involving patients with cancer to the in-depth analysis of the very early events which take place in the cells which go on to form tumours. The use of tissue culture has been indispensable in the latter course of study and it is this approach which will be presented here.

Normal Growth and Transformation

Normal growth and cancer are two faces of the same coin, and much of cancer research involves the study of controls on normal cell growth, in the hope that we will eventually understand how these normal controls have been disrupted in the cancer patient. In the normal cell, growth is controlled by a variety of positive and negative control pathways and these are finely balanced to induce growth only when required. An example of a positive control pathway which stimulates growth is the well controlled cascade of events that occurs when a mitogen binds to its receptor and induces cell proliferation. Negative control pathways exist to ensure that each step along this pathway is switched off when it is no longer needed. However, in this complicated scheme of stimulatory and inhibitory pathways there is much scope for errors to occur, and occasionally they do. The result is that cells lose their developmental controls and start to proliferate when they should remain quiescent. In other words, cancer can be viewed as a deregulation of the fine balance between positive and negative growth control mechanisms. This loss of growth control is associated with a number of measurable changes in the phenotype of cells and when a cell acquires these phenotypic properties it is said to be transformed.

'Normal' and Non-transformed Cells

Research into the molecular basis of cancer depends heavily on cultured cells, although clearly this work has to be complemented by *in vivo* experiments. Cultured cells have the advantage that the environment of the cells can be manipulated by the investigator, the type of target cell can be well defined, and changes in cells following treatment with carcinogens can be investigated. Furthermore these cells can be manipulated genetically.

As stated already that most diploid cells have a limited life span, and only rodents routinely give rise to continuous cell lines from normal tissue. Although these cell lines maintain many of the characteristics of the cell type *in vivo*, they cannot really be described as normal for a number of reasons. These cells tend to divide faster than cells *in vivo*, they have a slightly reduced cell size, and more importantly, they are genetically different from the original cells, since the culture is aneuploid. However, if these cells are maintained at low cell density, they retain many of the characteristics of normal cells. It is useful at this stage to make the distinction between 'normal cells' and 'non-transformed cells'. The designation of cells as normal implies that growth control is normal, that the cells have undergone no genetic changes, and that the cells have not been transformed. By these standards, immortalized rodent cell lines are not normal. They are, however, very useful for the study of cellular transformation since many of these cell lines are non-transformed. Treatment of such adherent cells with various agents can dramatically change the growth properties of cells in culture: the cells become transformed.

Cell Transformation

Transformation is defined as the acquisition by cells of certain phenotypic properties associated with cancer. A number of assays have been developed in order to establish whether cells are transformed and they have generally proved reliable if used in the correct context.

Transformed adherent cells can be identified by changes in the control of their growth and behaviour, and some of these changes are described in more detail below.

Phenotypic Properties of Transformed Cells

The changes to cells when they become transformed can be treated under five headings:

- Ability to form foci in *in vitro* culture;
- Changes in morphology;
- Changes in growth factor requirements;
- Release from anchorage-dependence for cell division;
- Ability to form tumours *in vivo*.

These changes provide the basis for determining whether or not cells have been transformed.

Focus formation

The most obvious and medically significant property of cancer cells is that they proliferate uncontrollably. When normal cells are grown in culture, they form a monolayer and when this monolayer reaches confluence the cells stop dividing through a process called *contact inhibition*. Transformed cells loose this property, and produce disorganized clusters of cells which grow on top of each other. These are called 'foci' (singular, focus), and these are readily recognizable when seen against a background of normal cells growing as a monolayer. This assay is often referred to as the focus forming assay.

Morphology

The morphology of cells changes upon transformation. Usually transformed cells tend to round up, probably due to the disruption and reorganization of cytoskeletal structures.

Growth factor requirement

Transformed cells proliferate under conditions in which normal cells do not. When such cells are cultivated *in vitro* they will grow in media containing low serum concentrations whereas non-transformed cells will not do so. This reduced requirement for exogenous mitogenic stimuli is now known to result from abnormal expression of growth factors or growth factor receptors or by subversion of the signal transduction pathways involved in normal growth control.

Transformed cells become anchorage-independent

As discussed already that most cells (with the exception of haemopoietic cells) grow as an adherent monolayer. Adherent cells

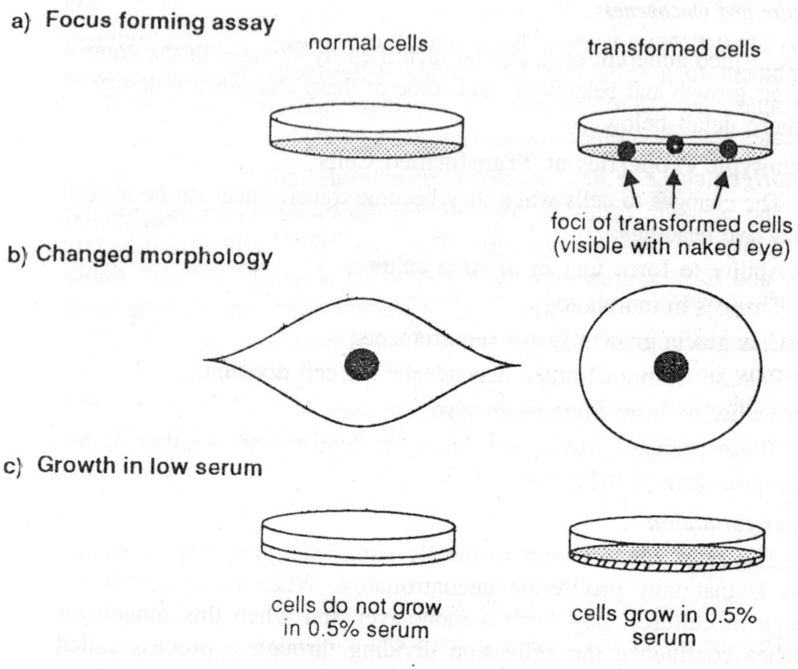

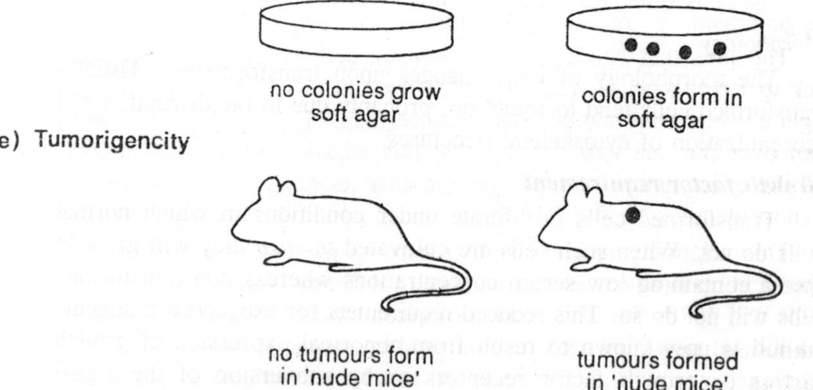

Fig. 15.1. Diagrammatic illustration of the techniques used to demonstrate the presence of transformed cells in a culture.

are said to be anchorage-dependent which means that they need to attach to a substrate in order to proliferate. When cells are transformed, they acquire anchorage-independence, the ability to grow without attachment to a substrate. In practice, when cells are suspended in soft agar, growth of colonies from single cells in such a medium indicates that the cells have become transformed.

Tumorigenicity can be demonstrated in 'nude' mice

The ultimate criterion for cell transformation is tumorigenicity: the ability to generate tumours. This can only be demonstrated *in vivo*, and is measured experimentally using a special kind of mouse called the '*nude*' mouse (also called the *athymic mouse*). The nude mouse is a developmental mutant which has lost its ability to produce a thymus gland and therefore has a defective immune defence system. It can therefore serve as a host for the growth of tumorigenic cells originating from any species, since these cells will not be recognized as foreign by the animal's immune response. The rate of tumour formation and the number of tumours formed when cells are injected under the skin of the immune deficient mouse gives an indication of tumorigenicity.

Cancer is Caused by Changes in the Genome

The phenotypic differences between non-transformed and transformed cells covered above provide only a descriptive account of transformation without explaining how or why transformation takes place. One clue as to the causes of transformation is the fact that these phenotypic changes are a reflection of genetic changes. The genetic material of a cell is somehow changed and cell populations with these genetic changes are generated by growth and division.

The idea that cancer may be essentially a genetic disease dates back to 1914, when Boveri suggested that chromosomal abnormalities might convert a normal cell into a cancer cell. However, it has largely been over the last two decades that this idea has been substantiated and the exact nature of these genetic changes has been characterized in more detail.

This relatively recent insight regarding the cause of cancer was dependent upon advances in a number of different branches of science. In the mid 1970s the development of recombinant DNA technology led to a radical new approach to biological research. Recombinant DNA techniques permit the formation of new combinations of genes by allowing us to isolate a DNA fragment, change it in a defined way and introduce it into the same organism or a different organism. Thus it became theoretically possible to isolate genes responsible for transformation and study them in detail. The study of viruses actually led to the discovery of such genes. The first gene identified which could transform a normal cell, causing the phenotypic changes associated with cancer, was isolated from a virus. Subsequently such genes, called oncogenes, were isolated from other sources including human tumours. Further study indicated that oncogenes were mutated versions of normal cellular genes which playa role in normal growth control. The normal genes were therefore called *proto-oncogenes*. In summary, proto-oncogenes playa role in normal growth control but certain mutations of such genes result in subversion of this growth control leading to cancer.

It is important to remember that cancer reflects changes in the genome (i.e. it may be regarded as a genetic disease). It is not necessarily inherited. In fact, inherited forms of cancer are rare, although many cancers do show familial clustering.

A genetic disease involves changes in the genetic material of the cell, in other words, in the cell's DNA. When such a change (a mutation) occurs in the somatic cells of the body, it will not be transmitted to subsequent generations. Only mutations in the germ-line cells will result in the mutated genes being present in all the cells of the offspring and will be carried to each subsequent generation. In an inherited disease the mutation is present in the germ-line cells and therefore passed on to the next generation. Thus cancer is a genetic disease but not necessarily an inherited disease.

MOLECULAR BASIS OF CANCER

A very important goal in the field of cancer research is the identification of genes that convert a normal cell into a tumorigenic cell and the genes that suppress the phenotypic properties of a tumorigenic cell. Molecular biology has opened the door towards the definition of three major groups of genes involved in neoplasia. These are:

Oncogenes which act dominantly in the sense that they are able to induce at least some of the changes in the cell phenotype associated with the neoplastic state when transferred into cultured cells. An oncogene is defined as a gene that contributes to neoplastic transformation when it is introduced into a non-transformed cell;

Tumour suppressor genes which have to be deleted or functionally incapacitated before a tumour can arise;

Modulators which modify the spread of neoplastic cells in the organism, for example, by influencing metastatic spread, invasiveness, or cellular resistance to immune rejection. Many potential cancer cells are prevented from causing neoplasia by the immune system, which recognizes and destroys the transformed cells. The importance of immune surveillance in keeping potential tumours at bay is demonstrated in patients who are immunodeficient. In these patients the eventual cause of death frequently involves tumour growth.

ONCOGENES

In 1911, Peyton Rous demonstrated that cell-free filtrates of certain chicken sarcomas (tumours of connective tissue) promote new sarcomas in chickens. Interestingly, Peyton Rous was one of many people who did not believe that genetic mutations played a role in the genesis of cancer, but the virus he discovered eventually led to the proof that genetic changes can cause cancer. The virus which had caused these tumours was later identified as *Rous Sarcoma Virus* (RSV). RSV is a retrovirus (an RNA virus that replicates itself by copying its RNA into DNA, inserting this DNA into the host cell genome, and then transcribing the DNA). The first oncogene to be discovered, *v-src,* was identified from RSV, and its protein product, a 60 kD phosphoprotein (called $pp60^{v\text{-}src}$) was found to be responsible for the induction of tumours and neoplastic transformations by RSV.

Discovery of Proto-oncogenes

In 1976, Michael Bishop and Harold Varmus made the remarkable discovery that uninfected chicken cells contain a gene which is homologous to viral *v-src*. This cellular gene was called *c-src*. Moreover, species as divergent as humans and the fruit fly *Drosophila* all contained a similar gene. This raised the question 'what were these potential cancer-inducing genes doing in the normal genome?' We will attempt to answer this.

Role of Proto-oncogenes

In 1983, there was a breakthrough in the understanding of the role of proto-oncogenes. By this time a number of different oncogenes and proto-oncogenes had been identified and one oncogene, called *v-sis* (found in Simian Sarcoma Virus, SSV for short) which had been sequenced, was found by computer searches to be homologous to the gene encoding one of the chains of human *platelet-derived growth factor* (PDGF). In other words, the gene coding for this PDGF chain was the proto-oncogene corresponding to the *v-sis* oncogene. This suggested that proto-oncogenes are genes encoding proteins which playa role in normal growth regulation. A non-transforming ancestor of SSV had picked up, or hijacked, the normal PDGF gene from the cellular genome by a

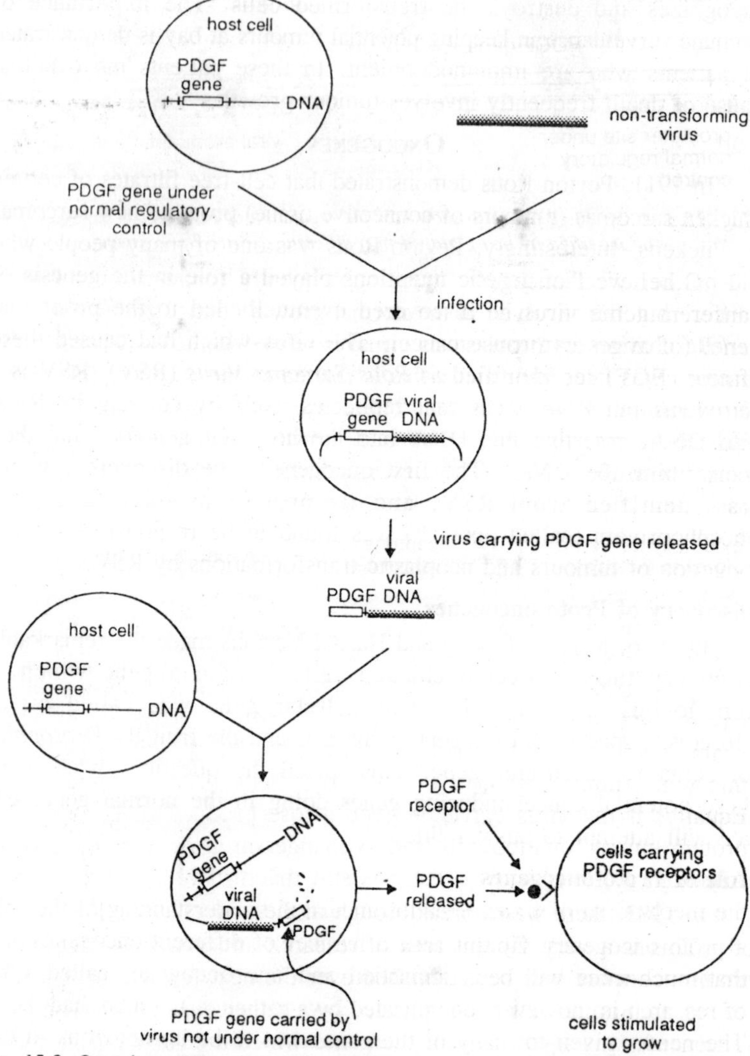

Fig. 15.2. Overview of the process by which a non-transforming ancestor of SSV may have picked up the PDGF gene to become a transforming virus.

process known as *genetic transduction*. The PDGF gene, now in a different genetic 'environment' was not under the normal regulatory control. Thus, when the PDGF gene-carrying virus infected a host cell, the PDGF gene was expressed and thus any cells with PDGF receptors were stimulated to grow.

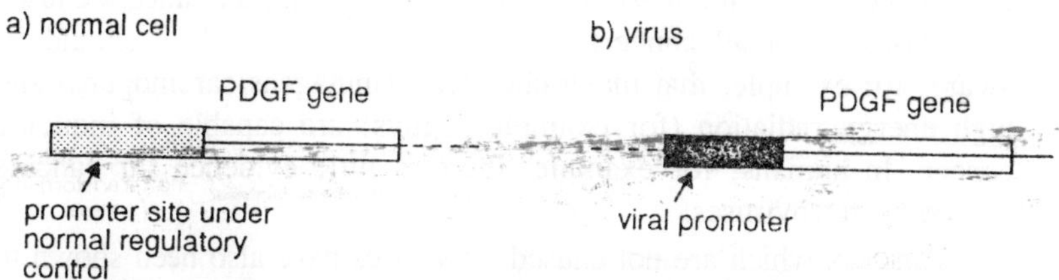

Fig. 15.3. A simplified figure to show how the regulation of PDGF gene expression may be modified by being placed in a different genetic environment.

Other viral oncogenes were subsequently found to be related to different genes involved in the normal growth control pathway e.g., *v-erbB* (of avian erythroblastosis virus) is related to the epidermal growth factor (EGF) receptor and *v-fms* is related to the receptor for *colony-stimulating factor-l* (CSF-l). There are many more examples. As indicated before that the regulation of growth and division of mammalian cells is complex. We can, however, attempt to represent this regulation as a simple cascade in the following way:

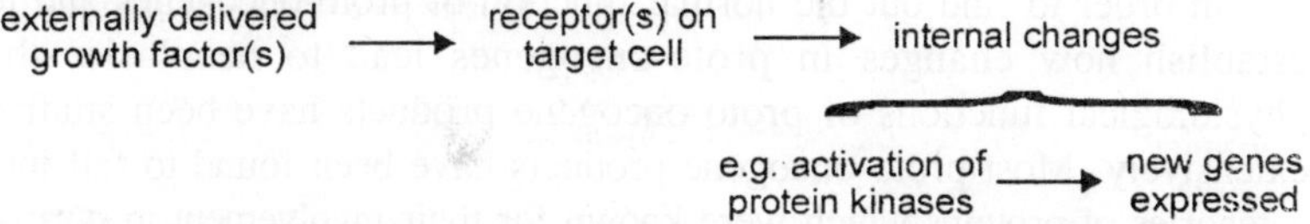

In principle any gene carried by a virus which bypasses or changes any of these steps will lead to changed growth characteristics. For example if the gene carried by the virus produces a growth factor, this will stimulate growth of the target cells of the growth factor. Equally, if the virus carries a gene whose product activates the key protein kinase, this too will lead to changes in growth characteristics. Although much is known, there is still much to be done to establish the mechanism of transformation mediated by many viruses.

This is a very vibrant area of research and you should anticipate that much more will be established over the next few years. This area of research is, however, complicated by a rather difficult nomenclature. The names given to many of the genes which are involved in normal growth control are confusing since many of these genes were discovered by studying systems in which they had gone wrong. For example, the *c-src* gene is involved in normal growth control but owes its name to the sarcomas that the oncogene *v-src* causes in chickens.

Are all Oncogenes Viral?

In the previous section it was told that the oncogenes were discovered in retroviruses and that these genes were derived from host proto-oncogenes. The question we might ask is 'are all cancers caused by viruses?' Not all cancers are caused by viruses. You should be aware, for example, that many chemicals (mutagens/carcinogens) and high energy radiation (for example X-rays) are capable of inducing cancer. In humans, for example, there is little evidence for cancers caused by retroviruses.

Tumours which are not caused by viruses have also been shown to contain oncogenes by a technique known as transfection. Most non-viral oncogenes were identified by transfection of DNA from human cancer cells into mouse 3T3 cells. This technique results in transformation of cells, and the gene which caused the transformation can subsequently be isolated and studied in detail. One group of oncogenes, called *ras* genes, has been found in 30% of human cancers. Like the viral oncogenes, these oncogenes were also found to have homologous proto-oncogenes in normal cells only these proto-oncogenes were activated into oncogenes by different mechanisms.

Classification of Oncogenes and Proto-oncogenes

In order to find out the normal function of proto-oncogenes and to establish how changes in proto-oncogenes lead to neoplasia, the physiological functions of proto-oncogene products have been studied extensively. Most proto-oncogene products have been found to fall into categories of proteins which were known for their involvement in normal growth control. Some proto-oncogene products resemble known growth factors, others code for growth factor receptors, yet others for intracellular signal transducers, and finally some resemble nuclear transcription factors. This finding exemplifies how important it is to study normal growth as well as the events which occur upon transformation.

Mechanisms of Activation of Proto-oncogenes giving rise to Oncogenes

Proto-oncogenes can be activated by a number of different mechanisms, some of these have already been mentioned. Here they are grouped together into different classes.

Changes which result in an altered protein

Even single point mutation can render a gene oncogenic by changing the structure and function of the protein which it codes for. For example the *ras* proto-oncogene which was mentioned above is often activated into an oncogene by a single point mutation. In this particular example the product of the gene sends signals instructing the cell to divide in an unregulated manner so that the cell divides when it should not. Usually a point mutation have to occur in order for a proto-oncogene to become an oncogene. A point mutation within the structural gene may lead to the production of a protein with changed properties. For example it might bind more, or less, strongly with its receptor. Alternatively if the protein is a receptor for a growth factor, it might be able to activate the next step in the control cascade even in the absence of the growth factor.

We can represent this in the following way:

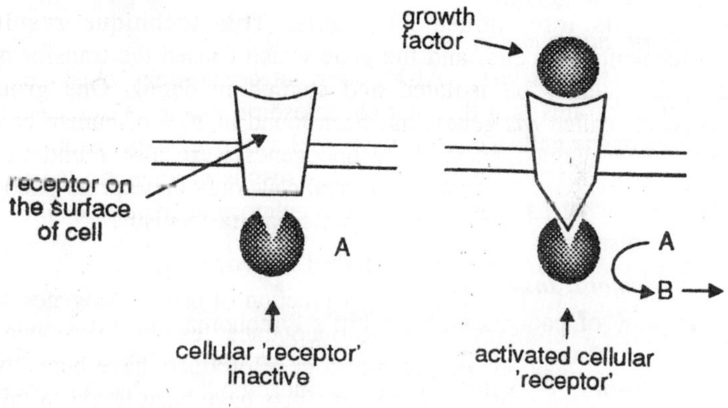

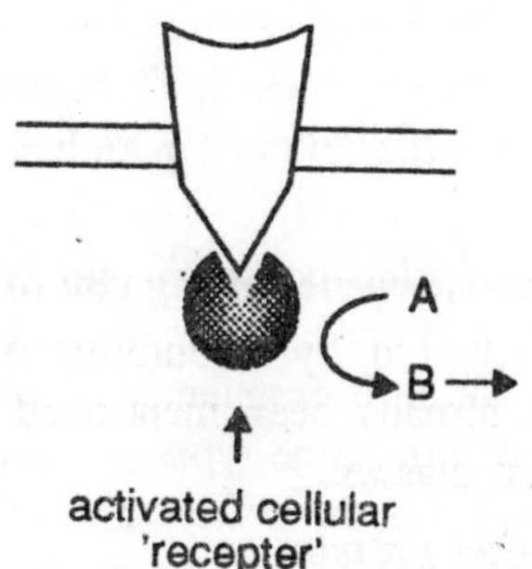

In the normal cell (a), events in the cell are regulated by the presence/absence of the externa l growth factor. In (b) the 'modified' receptor activates cellular processes even in the absence of the growth factor. If the point mutation occurs in the regulatory sequence of a key proto-oncogene, it may be that the expression of this gene is no longer sensitive to normal control. Thus the gene may become over- or under-expressed. This may have quite a profound effect on the growth of the cell. For example, we might get over-expression of a gene coding for a growth factor or a growth factor receptor. It might also lead to the expression of, for example, a growth factor receptor in the wrong cells. The recently developed PCR (*polymerase chain reaction*) technique is ideal for the study of small changes in the DNA of human tumours since it allows the expansion and analysis of small sections of the genome.

Insertional mutagenesis

Avian leukosis virus (ALV) does not contain any oncogenes, but when it integrates into the host chromosome it often inserts near to a proto-oncogene called *c-myc*. This results in over-expression of the *c-myc* gene product in the nucleus which results in uncontrolled growth. Effectively what happens is that the regulatory sequences of the *c-myc* gene are disrupted.

Chromosomal translocation

In 90% of patients with Burkitt's lymphoma, the proto-oncogene *c-myc* is activated by translocation from its normal position at the end of chromosome 8 to the end of chromosome 14, adjacent to certain immunoglobulin genes. This brings *c-myc* under the transcriptional control of the immunoglobulin regulatory sequences, which results in over-expression of the *c-myc* protein in lymphoid cells. In addition the immunoglobulin genes undergo germline rearrangement in order to generate antibody diversity and the location of *c-myc* in this region of the genome results in a high somatic mutation rate of the oncogene.

DNA amplification

Amplification of large parts of chromosomes can be observed cytogenetically whereas amplification of specific genes can be observed by Southern blotting. Gene amplification is frequently accompanied by over-expression of the gene product. For some types of cancer there is a correlation between poor prognosis for the patient and amplification of specific oncogenes. The finding that many copies leads to poor prognosis is in itself a useful discovery since it helps to decide the treatment for patients with cancer, and eliminates unnecessary treatment for those with good prognosis.

Mutations in the regulatory sequences of genes can also result in over-expression of the gene product, with similar results as gene amplification. It should be self evident that any agent that might bring about changes in DNA sequences is potentially able to convert proto-oncogenes into oncogenes. The mechanisms described above explain why chemicals which cause nucleotides to be inserted in the wrong place in DNA are potentially carcinogenic. The excision from and replacement of nucleotides in DNA damaged by high energy radiation is also error prone. The association of cancers with nuclear radiation can also be accounted for by the mechanisms described above.

TUMOUR SUPPRESSOR GENES

A tacit assumption in the early studies of cancer was the dominant nature of activated oncogenes. However, earlier studies had clearly shown that inactivation of tumour suppressor genes could also playa role in transformation. It is harder to identify tumour suppressor genes than oncogenes because of the very nature of their function. An oncogene contributes to neoplastic transformation when it is introduced into a non-neoplastic cell. It is easy to spot the change when one cell starts to divide in an uncontrolled manner, since visible clumps of cells form. Imagine what would happen if one cell acquired a tumour suppressor gene. The ability to cause reversal of uncontrolled growth would be much harder to observe. Our understanding of tumour suppressor genes, like our understanding of oncogenes, is heavily dependent on the use of cell culture models.

Somatic Cell Hybrid Studies Reveal Tumour Suppressor Genes

In 1969, Harris and Klein demonstrated that when malignant cells are fused with normal cells the resulting hybrid cells were non-tumorigenic. They reasoned that this tumour suppression indicated that genes from a normal cell might replace a defective function in the cancer cell and render it 'normal'. Such genetic elements have been called *tumour suppressor genes*. These findings were controversial for a while, because other researchers reported the opposite: that the hybrids were *tumorigenic*. It is now understood that their findings were due to the chromosomal instability of the hybrid cells. Chromosomes are rapidly lost from these hybrid cells, and the tumorigenic segregants have a positive growth advantage over other cells and are therefore selected for. However, when these technical problems were overcome, it was clear that many types of tumour do arise through loss of critical growth regulating genes from the genome.

Clinical Studies

Further evidence for tumour suppressor genes came from clinical studies of inherited cancer susceptibilities. The best studied example of inherited cancer is retinoblastoma, where a loss of gene function appears to playa role. Children who inherit one defective copy of the *Rb-1* gene develop, on average, 3 retinoblastoma tumours each. In 1971 Knudson produced a theory which he called the '*two-hit*' theory in which he proposed that all types of retinoblastoma involve 2 separate mutations. In sporadic retinoblastoma both mutations occur somatically in the same retinal precursor cell, whereas in heritable retinoblastoma, one of the mutations is germinal and the second is somatic. So in inherited retinoblastoma, the first event is the acquisition of the *Rb-1* defect before birth. The second event is the deletion or mutation of the remaining *Rb-1* gene.

Molecular Genetics Provides the Ultimate Proof of the Existence of Tumour Suppressor Genes

The ultimate proof that deletion of a tumour suppressor gene plays a role in tumour progression must involve the identification of the genes involved. Introduction of the gene into a cancer cell that lacks a functional copy should restore the phenotype to normal. In practice the phenotype of cells derived from tumours can rarely be expected to revert to a completely normal phenotype, since a large number of changes may have occurred in the genome. Only a reduction in the malignant phenotype can be expected by correcting one defect.

Since the development of recombinant DNA techniques, molecular geneticist have focused their attention on malignancies with defined chromosomal abnormalities. Molecular cloning of tumour suppressor genes is very difficult and therefore the list of tumour suppressor genes which have been identified is inevitably much shorter than the list of oncogenes.

As mentioned already, one tumour suppressor gene which has been identified is the *Rb-1* gene which is deleted in retinoblastoma. The infection of *Rb-1* deficient retinoblastoma cells in culture with a retroviral construct containing a full length *Rb-l* cDNA provides the final proof that *Rb-1* is really a tumour suppressor gene. The outcome shows that expression of the *Rb-1* gene product results in growth suppression *in vitro*.

Multi-stage Carcinogenesis

There is now considerable evidence that the activation of a cellular proto-oncogene to an oncogene and the deletion of tumour suppressor genes can both be causal events in carcinogenesis. However, single oncogenes under most circumstances are insufficient to cause full transformation. Instead transformation apparently requires a cell to have undergone multiple genetic lesions, which include both activation of oncogenes and inactivation of tumour suppressor genes. It has been thought for many years that the development of cancers takes place in many stages. The current theory of carcinogenesis is that it is a multi-stage process. The activation of oncogenes and inactivation of tumour suppressor genes can cooperate, resulting in fully transformed cells.

1. UV light causes adjacent thymidine residues on DNA strands to form dimers. The cell's DNA repair mechanisms excise these dimers and replace them. Sometimes incorrect nucleotides are inserted. Use this process to explain why individuals exposed to high intensity sunlight over long periods often develop skin cancers.
2. Try to describe a mechanism by which exposure to UV radiation might give rise to growth abnormalities in organs other than skin.

Importance of Cancer Research to Biotechnology

It should be clear to you that the techniques of contemporary biotechnology have made important contributions to our understanding of the mechanisms of cancer induction. The ability to identify, clone and sequence genes has, for example, enabled us to identify and compare proto-oncogenes and oncogenes. This, however, is not a one way process. Cancer research has provided a focus for the application of biotechnological techniques and has provided an important impetus to the development of biotechnological expertise. In addition, discoveries within the context of cancer research have contributed significantly to our understanding of how growth and gene expression is regulated in normal cells. This increased understanding has potential applications leading to improved cultivation of animal cells *in vitro*. The *in vitro* cultivation of animal cells for use in toxicity and drug efficiency testing and in the production of vaccines, antibodies, growth factors, hormones etc are important commercial processes. Improvements in the cultivation of animal cells, derived from the discoveries made in cancer research, could, therefore, have important consequences to the commercial success of these operations.

The discoveries made in cancer research have another important bearing on the biotechnological industries by providing new targets for the application of biotechnology. By understanding how cancers develop, it becomes possible to identify new approaches to treating these conditions. The production of targeted drug delivery systems to treat cancers depends upon harnessing biotechnological processes. Similarly cancer research may enable us to identify potential 'gene therapy' approaches to solving problems associated with some cancers. For example, we described the high incidence of retinoblastoma in individuals who carry a defective copy of the tumour suppressor gene *Rb-1*. This raises the possibility of using a genetic approach to reducing the incidence/consequences of this defect. Such an approach would, of course, be dependent upon the application of biotechnological techniques.

Another important area of biotechnology is in the production of diagnostics. The identification of, for example, transformed cells is important in cancer diagnosis and treatment. Again, the biotechnology industry has the potential to generate appropriate diagnostic reagents. In the same context, diagnostic reagents can be used for embryo screening to identify embryos carrying genes which make them susceptible to generating particular cancers. Biotechnology has the potential to generate, for example specific gene probes which enables us to identify if particular genes are present in an embryo/foetus. For example, if a probe was produced which enabled us to identify a defective tumour suppressor gene *Rb-1* in an embryo at an early stage, it might facilitate treatment. There are, of course, many ethical issues associated with such courses of action which need to be resolved. We do not propose to discuss these at this stage since our objective is to illustrate how biotechnology has contributed to cancer research and how cancer research may contribute to the development of biotechnology.

16

Culture and Genetic Engineering

Genetic engineering is a relatively new methodology which was developed in the mid-1970s. At this time a succession of new methods evolved which facilitated the formation of new combinations of genes by allowing us to isolate DNA fragments, change them in defined ways, and re-introduce them into the same organism or to introduce them into a different organism. This is the essence of what has appropriately been called genetic engineering.

One aspect of genetic engineering is the *in vitro* manipulation of genes by the investigator. Gene manipulation has had to be defined legally as a result of government legislation to control it. The UK definition is 'the formation of new combinations of heritable material by the insertion of nucleic acid molecules produced by whatever means outside the cell, into any virus, bacterial plasmid, or other vector system, so as to allow their incorporation into a host organism in which they do not naturally occur, but in which they are capable of continued propagation'. Other national definitions may differ in the exact words used, but are very similar in sentiment.

The techniques needed for gene manipulation are called *recombinant DNA techniques*. Thus genetic engineering, recombinant DNA technology, and gene manipulation are related terms, and each term refers to a slightly different aspect of the overall picture, although the words are sometimes used interchangeably.

The genetic engineering of animal cells in culture has many possible applications. In general, genetic material is introduced into cells so that they will synthesize a protein which they would not normally synthesize. The objective may be to analyze the effects of a specific gene within a cell, or to produce a protein which has been processed correctly so that the protein can be further studied, or to correct a genetic defect in a human patient. The ability to introduce cloned genes into cells is a very important and powerful method, playing a critical role in many areas of biotechnology and medicine.

Why Choose Animal Cells to Produce Specific Proteins?

Although bacteria are in many ways easier to work with to produce proteins on a large scale, the main problem with this system is that the genetically engineered products of bacterial cells are not necessarily identical to the products of the same gene when expressed in eukaryotic cells. One of the main obstacles is that bacterial cells process proteins in a different way from eukaryotic cells. To become mature proteins, polypeptides must fold into their native conformations and any subunits must properly combine.

Proteolytic cleavage may be necessary and there is a number of covalent modifications which may need to take place for example, glycosylation (additional of carbohydrate residues), acetylation, hydroxylation, methylation, nucleotidylation, phosphorylation, ADP ribosylation, and formation of disulphide bonds. Most of these modifications do not playa role in bacterial proteins, and so the relevant enzymes are generally missing in bacteria. Many of these post-translational modifications do not even occur in yeast or other lower eukaryotes and animal cells may have to be used to produce authentic proteins. In addition, foreign proteins over-expressed in bacteria often form clumps of denatured proteins and many foreign proteins are broken down by bacteria. These problems do not generally occur in animal cells, and furthermore animal cells can be made to secrete their products into the cell culture medium, making purification easier.

There are, of course, also many disadvantages of working with animal cells. They are more fragile, slower to grow, have more complex nutrient requirements, and great care is needed to avoid contamination of cultures.

It is worth mentioning that even if animal cells are used there may be problems in getting the correctly processed protein product. This is because there are differences in the ability to glycosylate proteins (in other words the ability to add carbohydrate residues to the side-chains of amino acids) depending on the species, the cell type, and the culture conditions. Even the process of transfection can induce the expression of previously unexpressed glycosyl transferases (the enzyme which carries out glycosylation). Thus it is essential to assess the glycosylation pattern of glycoproteins, in particular when they are destined for therapeutic use.

What Kinds of Products are Made using Animal Cells Cultured *In vitro*?

As mentioned earlier that animal cells have been used commercially in the production of viruses for vaccines, and for the production of monoclonal antibodies. It was not long before pharmaceutical companies became interested in generating proteins of therapeutic value from genetically engineered cells. The manufacture of proteins from animal cells has become an important component of the biotechnology industry. Examples of recombinant proteins which are now produced on a large scale from mammalian cells in culture include tissue plasminogen activator (used in the treatment of thrombosis), the blood clotting factors such as factor VIII (used in the treatment of haemophilia) and factor IX (used in the treatment of Christmas disease), and erythropoietin. In addition, genetic engineering of cultured cells has been important in generating new vaccines which are safer, have reduced side effects, and are more effective immunologically. The key to these improved vaccines was to produce only a viral protein rather than the whole virus, by expressing a viral gene product in cultured cells. In 1990 a hepatitis B vaccine was produced in this way using Chinese hamster ovary cells, and vaccines against influenza virus and HIV are in the experimental stages. Many of these products had previously been obtained from human blood provided by donors. However, it is often expensive to purify proteins from human blood, and there are risks of hepatitis and AIDS inherent in these products.

The genetic engineering of animal cells to produce recombinant proteins was dependent upon the development of recombinant DNA techniques—the ability to form different combinations of genetic material *in vitro*.

The advances in genetic engineering were also dependent upon the technology for introducing the manipulated genetic material into animal cells. This technology originated from the observations that naked uncoated protein-free viral DNA can enter cells and change the phenotype of cultured cells. Such transfer of genetic material was first achieved by Szybaska and Szybaski in 1962, although they were unable to demonstrate that cells had actually taken up DNA. Burnett and Harrington were the first to prove that cultured cells would take up DNA by treatment of cells with DEAE-dextran and viral DNA in 1968. The viral life cycle was initiated in a small proportion of these cells, and the cells were said to be transfected. The uptake of foreign DNA by animal cells has been called *transformation*.

Gene Manipulation

As stated earlier that gene manipulation involves making changes in the DNA, and a number of basic skills are needed to do this. In this method fragments of DNA are inserted into a vector to produce a recombinant DNA molecule. The vector acts as a vehicle which transports the gene of interest into the host cell. The host cell divides and copies of the foreign DNA are produced making a clone of cells carrying the foreign gene.

The relevant techniques include:

- Preparation of pure samples of DNA;
- Cutting of DNA;
- Analysis of DNA fragment sizes;
- Joining together of DNA molecules;
- Introduction of DNA into bacteria for amplification;
- Identification of bacteria that contain the correct recombinant DNA molecule.

Isolation of the Gene

In order to genetically engineer cultured cells, the gene of interest has first to be isolated, and this is not an easy task. For now, it is sufficient to know that a gene is frequently introduced as copy (complementary) DNA (cDNA) rather than as genomic DNA..

In eukaryotes, gene expression involves transcription, RNA processing, and translation, and that one of the events during RNA processing is the removal of introns from the heteronuclear RNA, which gives rise to mRNA. The mRNA can be used by the gene manipulator to create DNA by virtue of complementary base pairing, and this DNA is called *cDNA* (copy DNA). To achieve this we use an enzyme called reverse transcriptase. This synthesizes single stranded DNA using RNA as a template. Strictly speaking, this enzyme is called RNA-dependent DNA polymerase. More usually, however, it is referred to as reverse transcriptase. A double stranded DNA molecule may be synthesized from this single stranded DNA by using suitable DNA-dependent DNA polymerase. The main difference between genomic DNA and cDNA is that cDNA has no introns, whereas genomic DNA contains introns. Some genes contain so many introns that they become unmanageable and are too large to incorporate successfully into a vector, whereas the corresponding cDNA is less problematic. By using cDNA, the cell which receives this DNA by transfection does not have to 'process' (i.e. remove introns) it in order to produce mRNA.

Expression of the Gene

In order to produce a desired protein in animal cells in culture, it is important to introduce the DNA in a form that can be expressed. Gene expression refers to the synthesis of the final product of the gene. This product may be RNA, a protein, or a modified protein such as a glycoprotein. In the case of a protein product, the DNA has to be transcribed, and the transcripts have to be correctly processed, translocated to the cytosol, and subsequently translated. The polypeptide chain may need further processing, and the final product may either remain in the cell or be secreted. To optimize protein expression, all of these processes need to be considered. Apart from these requirements, the precise system used to express genetic information will vary with each task, and a number of choices regarding the type of expression system have to be made.

Stable or Transient Expression

A number of alternative techniques have been developed for introducing DNA into cells, and they fall into two groups:

(i) Those designed for transient expression;

(ii) Those in which stably transfected cell lines are selected.

In transient expression systems, DNA is introduced into cells and after 1-3 days the cells are harvested and the expression products of the introduced genes are analyzed. Since the DNA has not necessarily integrated into the genome, the long-term culture of these cells results in dilution and eventual loss of the gene of interest from the cells. Transient expression is often used in research but is less relevant to the large scale production of specific proteins. One area of research, which is worth mentioning, is the isolation of eukaryotic genes using transient expression systems.

The main aim of such experiments is to isolate a specific gene from the total DNA of a cell. This is quite a difficult task and there are a number of approaches depending on the type of information available about the gene of interest and the protein which it codes for. If antibodies have been raised to the protein, or if the protein has suitable characteristics, then the transient cloning expression system is often used. Genomic fragments, or cDNAs, are incorporated into expression vectors which are subsequently introduced into cells in culture. The transiently transfected cells are screened using antibodies to detect the protein of interest, and only those cells which have taken up the expression vector carrying the right gene will show a reaction with the antibody. The gene can then be isolated and studied in more detail.

In stable systems a small proportion of the cells will retain the genetic information, either because it has integrated into the chromosomes, or because sequences facilitating DNA replication are included in the construct (a piece of DNA constructed by recombinant DNA technology). The newly introduced DNA is passed on to all subsequent generations. Since the uptake of DNA into animal cells lines is not always very efficient, it is important to select those cells which have incorporated the foreign DNA.

Viral or Bacterial Plasmid Vectors

Ideally, DNA would be introduced into cells gently and efficiently without disrupting normal cellular processes. DNA can be carried into cells using vectors (literally 'carriers'), although as we shall see later, no ideal vector exists and in practice compromises have to be made. The cloning 'vectors used today are generally derived from either bacterial plasmids or viruses. As well as carrying the DNA into the cell, vectors have been constructed to allow gene expression by including a region called the expression cassette.

Plasmids are small pieces of circular DNA which occur naturally in bacteria and replicate independently of the bacterial genome. Genes of interest can be cloned into a bacterial plasmid, and the bacterial plasmid can be reintroduced into bacteria, where it replicates producing large amounts of the plasmid of interest. The simplest cloning vectors are based on the bacterial plasmid pBR322, which has two important features.

Firstly, it carries genes coding for antibiotic resistance (amp^R and tet^R), which allow for easy identification of bacteria which have taken up the vector. These are examples of selectable markers. Secondly, it carries a bacterial origin of replication (*ori*) which allows the plasmid to replicate independently of the bacterial genome. pBR322 allows manipulations to be carried out in bacteria and facilitates amplification of the recombinant plasmid in bacteria. The main reasons why pBR322 is inadequate as a mammalian cloning vehicle are that the two features mentioned above, (the ability to select for plasmid uptake, and the ability of the plasmid to replicate), would not operate in mammalian cells. In addition, you may have mentioned a third valid point, that is, that DNA incorporated into pBR322 would not be expressed in mammalian cells because the machinery for gene expression in mammalian cells would not recognize the bacterial expression regions.

Mammalian Vectors

The simplest mammalian cloning vectors are based on pBR322, but with a few modifications.

Let us look first at the ability to select those cells which have taken up the plasmid. As we shall see later, some of the methods of introducing DNA into animal cells are not very efficient, and it is therefore essential to select those cells which have stably incorporated the foreign DNA. In the case of bacteria, the selection of those bacteria which had taken up vector DNA was based on the use of marker genes such as amp^R or tet^R, followed by growing the bacteria in the presence of the appropriate antibiotic (ampicillin or tetracyclin). A similar strategy is used in mammalian cells, except that the antibiotic G418 is used. Mammalian cells are naturally sensitive to an antibiotic called G418, although cells expressing the *'neo'* gene become resistant to this antibiotic. Cells which express the *neo* gene can inactivate G418.

The *neo* gene can be added to pBR322, as in the plasmid pSV2*neo*. Any cell which has taken up pSV2neo will be able to grow in medium containing G418. Similarly there is a gene called *'hph'* which enables mammalian cells to survive in hygromycin C, and may be used as an alternative to the *neo*/G418 system. Both G418 and hygromycin act by inhibiting eukaryotic translation. The drug selection protocol is followed by isolation and characterization of individual transfectant clones. This is a labour intensive programme, requiring several weeks of tissue culture. Despite this drawback, the technique is commonly used, since the result is a stable expression system which can be used indefinitely.

What about the ability of plasmids to replicate in mammalian cells? We said earlier that plasmid based vectors have a bacterial origin of replication but we have not said anything about an origin of replication that facilitates DNA replication in mammalian cells. If a cell line is to produce a recombinant protein in a stable manner, then the gene of interest must be replicated in animal cells.

Again you will need an adequate knowledge of eukaryotic genetics to answer this question. In fact, the origins of replication in animal cells are poorly understood, and a discrete region which directs DNA replication has not been isolated. There are two tactics which are used to overcome this problem. Most plasmid-based systems depend on the integration of the genetic material into the mammalian genome, so that it is replicated along with the rest of the genome.

An alternative strategy has been to use the origins of replication of viruses which infect animal cells. These have been well characterized and facilitate replication of DNA in animal cells. These viral origins of replication are incorporated into plasmid vectors which are stably maintained as extrachromosomal elements at 10-100 copies per cell without integration. Such vectors are called *shuttle vectors*, since they can replicate in bacteria and in animal cells, for example the plasmid contains the bovine papilloma virus origin of replication (for replication in mammalian cells) and the pBR322 origin of replication (for replication in bacterial cells). There are also vectors which contain the SV40 origin of replication, and these cells can only replicate in the presence of a protein called the large *T antigen*. The gene for this

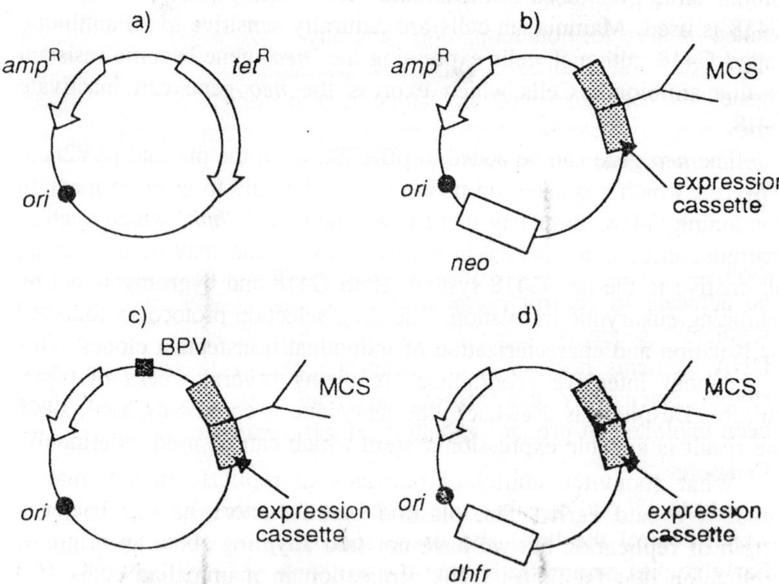

Fig. 16.1. Maps of pBR322 and various pBR322-derived eukaryotic cloning vectors. (a) pBR322—a commonly used bacterially derived plasmid. (b) pSV2neo is derived from pBR322 by addition of the neo gene which allows selection of transfectants using the antibiotic G418, and an expression cassette derived from the SV40 virus. (c) A shuttle vector containing the bovine papilloma virus and the pBR322. (d) pSV2dhfr is a pBR322-based amplification vector.

protein can either be provided on the vector, or it can be provided by the cells, as in COS cells (which are derived from African Green Monkey kidney). WOP cells, on the other hand, express polyoma virus T antigen and can be used in conjunction with expression vectors containing the polyoma origin of replication. Expression vectors based on adenovirus and vaccinia virus also exist.

Sometimes researchers want to have many copies of newly introduced DNA in order to obtain higher levels of gene expression, and a large number of amplification systems is now available. The system most commonly used is based on the *dhfr* gene coding for the enzyme dihydrofolate reductase (DHFR) which catalyses the conversion of dihydrofolate to tetrahydrofolate. The cytotoxic drug called methotrexate binds to the DHFR enzyme, but mammalian cells can develop resistance to it by amplifying the *dhfr* gene. More enzyme is produced so there will still be enough active enzyme after all the methotrexate is bound. After long-term culture of the cells in methotrexate, there may be up to 2000 copies of the *dhfr* gene per cell. Moreover gene amplification also occurs in cells which have taken up a vector containing *dhfr,* and any DNA integrated next to the *dhfr* gene will also be amplified. Expression vectors using the *dhfr* gene have been developed and can be used in many cells types, although they are particularly effective when used in conjunction with dhfr-deficient Chinese hamster ovary cell lines.

Expression Vectors

Most useful plasmid-based mammalian vectors contain an expression cassette. Whatever vector is used, the coding region of the gene must be adjacent to control regions which animal cells recognize. These regulatory sequences are different from those found in bacteria. The DNA elements necessary for eukaryotic gene expression have been studied extensively and information about these DNA elements has been exploited to produce a battery of gene transfer vectors.

Promoters

At the 5' end of eukaryotic genes are the promoters and enhancers which control the level and tissue specificity of DNA transcription. Early studies demonstrated that the structure of promoter elements is relatively conserved, although there are differences between the promoters used by different types of RNA polymerase. We will limit the discussion to the genes coding for proteins which are transcribed by RNA polymerase II (pol II). One critical aspect of all pol II promoter elements is that they contain mRNA cap sites—the point at which the mRNA transcript actually begins. By convention the transcription start site is + 1, and everything 5' or upstream of this is assigned a negative number, whereas everything 3' or downstream is assigned a positive number. Upstream of the cap site is a frequently encountered sequence known as the TAT A box. This centres around –25 (that is 25 base pairs upstream of the mRNA cap site) and is important for locating the start site of transcription. Further upstream, the relative positions of the promoter elements become more variable.

The CCAAT box resides roughly 70-80 base pairs upstream from the mRNA cap site and is important for the initial binding of RNA polymerase II. Another promoter element of many housekeeping genes consists of multiple copies of GC rich regions upstream of the mRNA cap site. There appears to be an absolute requirement for promoter elements in eukaryotic gene transcription, so they are included in all recombinant plasmids. In addition to promoter elements, other DNA consensus sequences are involved in the expression of genes in eukaryotes. They are called *enhancers* because they enhance the transcription of genes, and they were first identified in the genomes of viruses. An enhancer is a gene sequence that is required for the full activity of its associated promoter. It needs not have a fixed position or a fixed orientation relative to the gene. Enhancers are important for tissue specific gene transcription.

If the protein products have a harmful effect on the cell then there would be a selection against stable cell lines constitutively expressing such a product. However, using an inducible system, large amounts of cells can be grown up and then induced to produce the protein of interest, which can then be purified. A common approach has been the use of heat shock promoters such as the *Drosophila hsp* 70 promoter, which only directs expression of the adjacent coding region at 43°C.

Many different expression vectors have been developed using the coding portions of one gene and the regulatory portions of another gene. For example, a gene which is expressed at low levels can be combined with the promoter of a gene which is expressed at high levels in order to increase the level of expression of a protein. Many promoter regions have been taken from viruses. The retroviral *long terminal repeats* (LTRs), the SV40 early regions, and the adenovirus major late promoter are only a few examples of promoters which are frequently used in gene manipulation. At the 3' end of the gene, sequences necessary for termination of transcription, cleavage of the 3' end of RNA, and polyadenylation have to be present. Sequences

a) SV40-based expression cassette

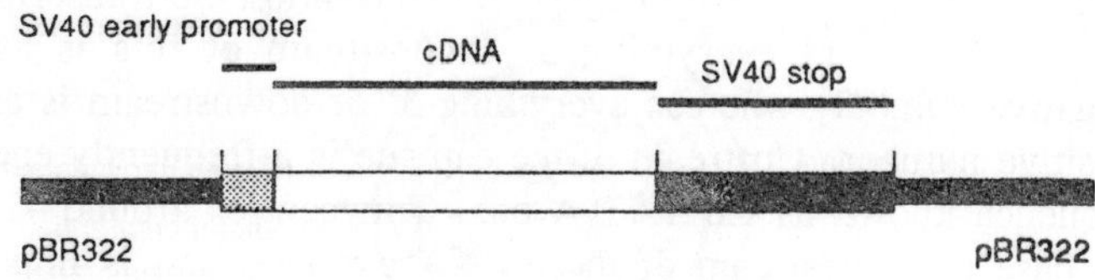

b) LTR-based expression cassette

Fig. 16.2. Diagram of 2 different types of expression cassettes.

which confer stability on the RNA transcript may be added to the recombinant plasmid. The recognition sequence for translation must also be present, and sequences which enhance the interaction of RNA with the ribosomal initiation complex may increase the rate of translation.

Retroviral Vectors

Alternatives to plasmid-based vectors are retroviral vectors. Retroviruses are RNA viruses that replicate by copying RNA to DNA inserting the DNA into the host cell genome, and then transcribing the DNA. Thus they have an RNA genome, but integrate a DNA copy of this RNA into the chromosomes of the cells they infect. Most retroviruses do not kill their host cells, but viral particles continually bud from the surface membrane of infected cells.

Retroviruses are very powerful tools in tissue culture since they have effective systems of infecting cells, they often contain powerful promoters, and their DNA integrates into the host cell genome as part of the normal retroviral replication cycle. These properties can be exploited by incorporating genes of interest into retroviruses and then using the genetically engineered retroviruses to infect cells. This is the most efficient method of introducing genes into cells and therefore deserves a more detailed description. We will start by giving some more information about the life cycle of the retrovirus.

Normal life cycle of a retr viruses

Retrovirally infected cells shed membrane encapsulated viruses. Each viral particle containing two molecules of RNA. The virus,

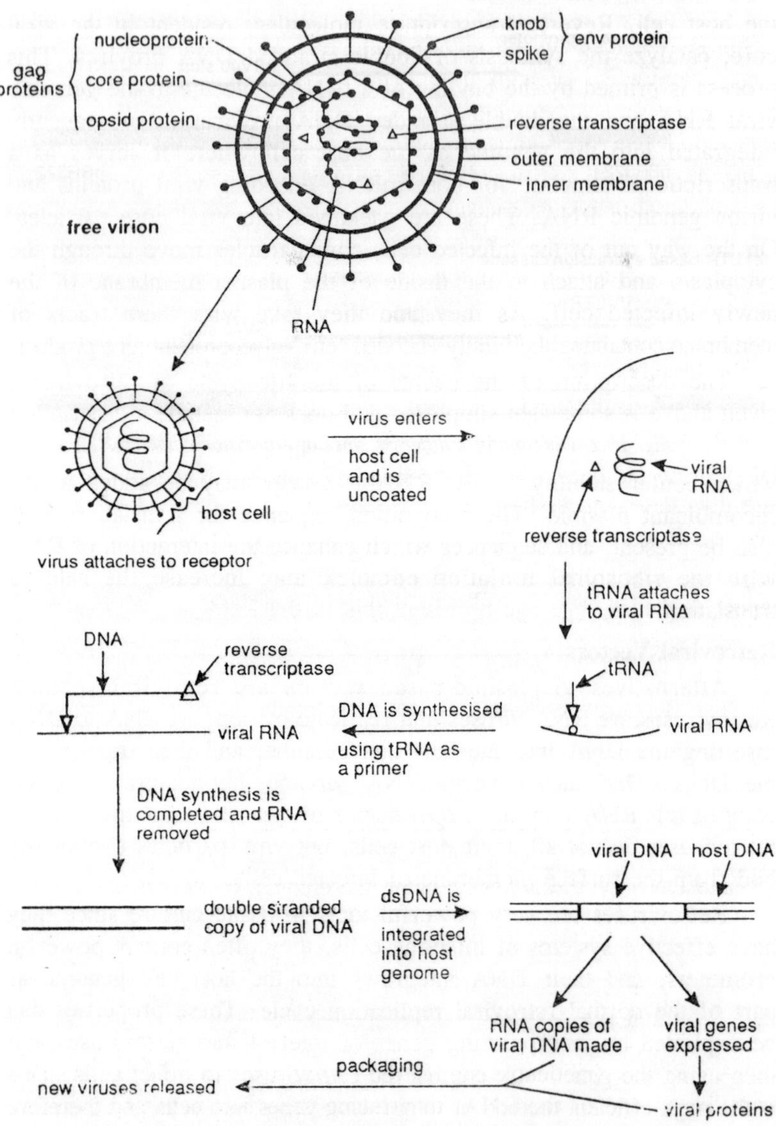

Fig. 16.3. Outline of the life cycle of retroviruses.

studded with an envelope glycoprotein (which serves to determine the host range of infectivity), attaches to a cellular receptor in the plasma membrane of the cell being infected. After receptor binding, the virus is internalized and uncoated as it is passed through the cytoplasm of the host cell. Reverse transcriptase molecules, resident in the viral core, catalyze the synthesis of double stranded DNA provirus. This process is primed by the binding of a tRNA molecule to the genomic viral RNA and the double stranded DNA provirus is subsequently integrated into the genome of the host cell where it serves as a transcriptional template for both mRNA encoding viral proteins and virion genomic RNA. These are packaged into viral core particles. On the way out of the infected cell, core particles move through the cytoplasm and attach to the inside of the plasma membrane of the newly infected cell. As they bud they take with them tracts of membrane containing the virally encoded 'env' glycoprotein gene product.

The description of the events of the life cycle of retroviruses given above is somewhat simplified and does not include a description of the events surrounding the synthesis and integration of double stranded DNA using the viral RNA as a template. This is a fascinating process and well worth discussion.

The RNA molecules of the virus contain terminal nucleotide sequences that are repeats of each other. They also contain an inverted repeat sequence. We can represent this in the following way:

3′ poly A ABC ———— C′B′A′ABC 5′

The tRNA attaches to this RNA molecule.

tRNA

3′ poly A ABC ———— C′B′A′ABC 5′

and DNA synthesis begins using the enzyme reverse transcriptase.

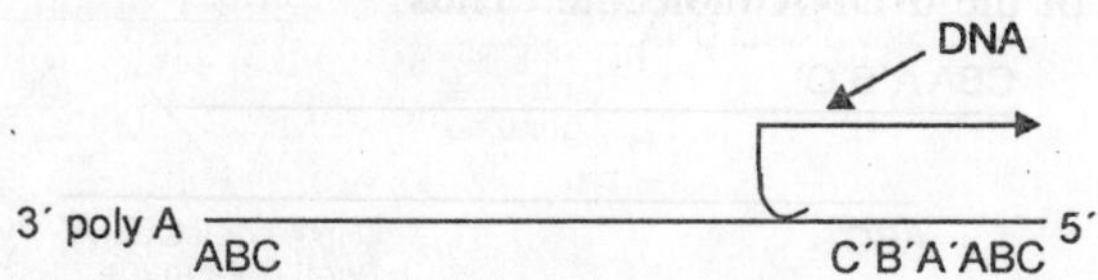

An RNase (ribonuclease H) then hydrolyses a few nucleotides from the 5′ end of the RNA, to produce the structure shown below:

CBAA′B′C′

3′ poly A ABC ———— C′B′A′ 5′

The second copy of the viral RNA then hybridizes (hydrogen bonds) with the DNA and DNA synthesis continues.

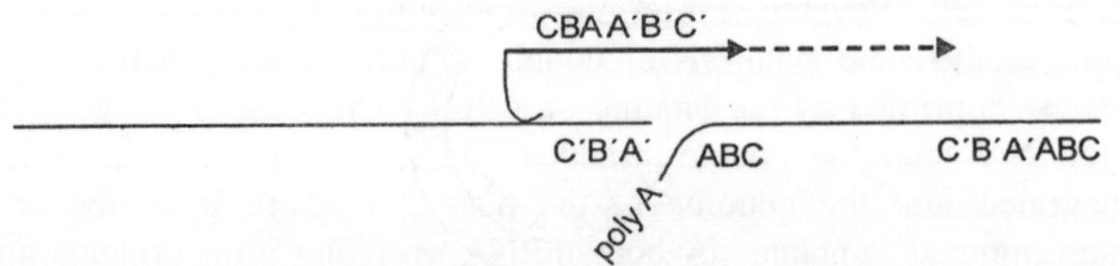

Ribonuclease H then digests away the RNA to leave ssDNA.

5' CBAA'B'C' CBAA'B'C' 3'

The 3' end of the molecule folds back on its self.

5' CBAA'B'C' CBA

C'B'A'

and DNA synthesis continues.

CBAA'B'C' CBA

C'B'A'ABC C'B'A'

An endonuclease cuts the loop to form a double stranded DNA molecule.

5' CBAA'B'C' CBA 3'

3' C'B'A'ABC C'B'A' 5'

The next question is, how does this DNA become integrated into the host genome?

It is thought that a 3' exonuclease digests a few nucleotides from either end of the dsDNA molecule. Thus:

The two ends of the molecule are now complementary so the molecule can then cyclise.

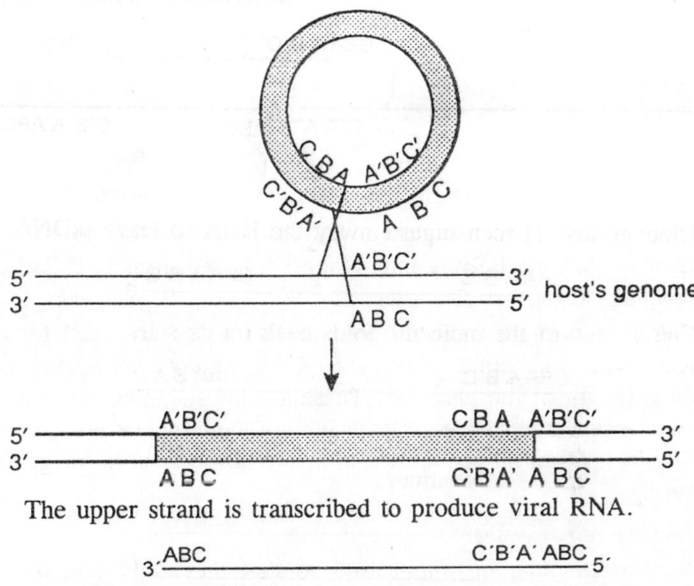

The upper strand is transcribed to produce viral RNA.

This elegant mechanism of viral replication and integration is well worth remembering. It is an excellent illustration of how repeat (and inverted repeat) sequences of nucleotides enable nature to manipulate nucleotides. It also provides a mechanism by which new genes (in this case retroviral genes) may be inserted and integrated into host genomes. In principle, we can hijack this system to introduce new genes into host cells.

The retroviral genome

In order to construct retroviral vectors we need to know about the sequences that are essential for integration and replication of the retroviral genome.

We anticipate that you would have decided that the terminal repeated sequences and inverted repeat sequence are essential. For simplicity, we will refer to them simply as *long terminal repeat* (LTR) sequences. You may also have included a binding site for the primer (tRNA). This is called the *Primer Binding Site* (PBS). You should have also predicted that the genome of the virus would need to code for the enzyme reverse transcriptase. This gene is usually called *pol* from polymerase.

There are, however, other genes. These include *gag* which codes for the viral core proteins and *env* which codes for the envelope proteins.

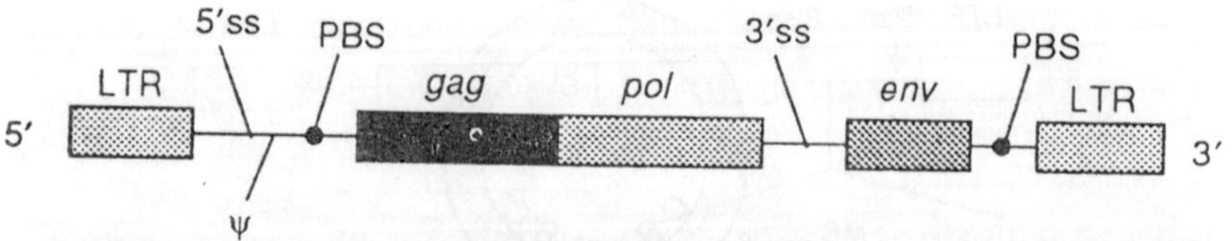

Fig. 16.4. Organization of the genome of a typical retrovirus.

There is also a short sequence which is necessary for the efficient packaging of the viral RNA into virions, this sequence is called the packaging signal (ψ).

There are two sites called 5'ss and 3'ss which are used for the production of sub-genomic mRNA (i.e. RNA that does not extend over the whole length of the genome). These are the so-called viral splice donor and acceptor sequences. Although all of these regions are essential for retroviruses, not all are needed to create useful vectors for genetic engineering.

Engineering retroviruses

Retroviruses can be engineered so that they only contain the sequences necessary for integration into the genome. The strategy used is to strip the virus of the dispensable *gag*, *pol* and *env* genes, thus generating a replication defective virus which requires other viral functions in order to replicate. The only things which are retained are:

1. The two LTRs (which facilitate integration);
2. The primer binding sites;
3. The packaging signal (ψ).

Many laboratories have shown that retroviruses containing foreign genes can be obtained in high titres. In 1982, a gene (the Herpes Simplex Virus thymidine kinase gene) was successfully integrated into the Moloney murine leukaemia provirus. Proviral DNA containing the thymidine kinase gene was transfected into host cells, but the virus could not reproduce itself in the normal manner since some of its essential functions were now missing (the virus was defective). These functions could be replaced using a helper virus, and the recombinant retroviral DNA was packaged into viral particles and secreted into the medium. The gene was expressed upon subsequent infection of cells with this recombinant retrovirus.

Retroviral vectors can be constructed using only the LTRs and immediate flanking sequences of the retroviral genome, and retroviral vectors with other useful features have been developed. In 1984 a

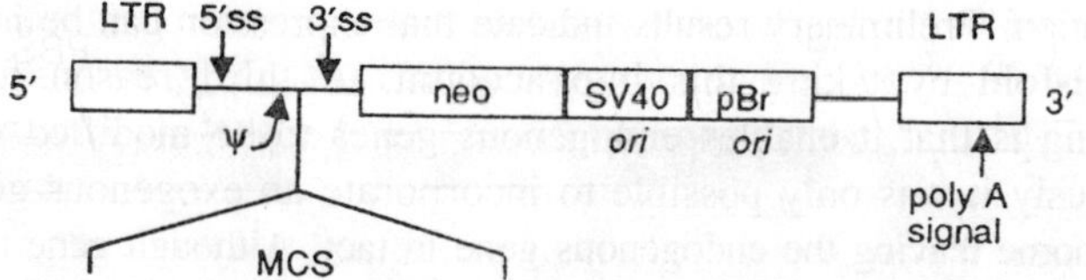

Fig. 16.5. Map of the recombinant retroviral vector pZIPNEOSV(X)1.

retroviral vector called pZIPNEOSV(X)1 was developed from a murine leukaemia virus and used to drive the expression of two inserted genes —the *neo* gene (included as a selectable marker) and the gene of choice. The expression of both genes can be driven from the 5' LTR. The vectors are used in combination with ϕ2 and ϕAM packaging lines. (Packaging lines are viruses that are used to co-infect cells and which will enable replication and packaging of recombinant retroviral vectors. For example if the vector carries no *pol* or *env* genes, it will only be replicated and packaged in cells containing viruses carrying these genes). Cells integrating the retrovirus can be selected using G418. MCS is a multiple cloning site, as it contains nucleotide sequences which allow us to use a variety of restriction enzymes to cut the vector in this region. New genes can then be inserted at this cut site.

Integration sites

Mammalian cells generally integrate newly introduced DNA in a non-specific manner. However, for several reasons it is desirable to introduce the gene at precise chromosomal locations.

The insertion of a DNA sequence into a specific site in the genome is often called *gene targeting*. More specifically the method used is called *homologous* recombination, since it invariably involves the combination of certain DNA sequences in the chromosome with homologous sequences present in the newly introduced DNA molecule. The opposite of gene targeting is random integration and one obvious disadvantage of random integration is that it may disrupt another important gene.

Another reason for gene targeting is that there seem to be certain locations in the genome which are expressed at higher levels than others and some regions are transcriptionally silent, although the reasons for this are poorly understood. For this reason the site of integration will affect the efficiency of recombinant protein production. Thus if the gene is integrated into a transcriptionally active site, the encoded protein may be expressed at higher levels than those obtained by random integration. Preliminary results indicate that expression can be increased several-fold by taking this into account. A third reason for gene targeting is that it enables endogenous genes to be modified whereas previously it was only possible to incorporate an exogenous gene into the genome leaving the endogenous gene in tact. Although gene targeting has been used for several years in yeast, it has only recently become possible in mammalian cells.

One of the largest areas of activity of gene targeting is in its application to the embryonic stem cells (ES-cells). The embryo is removed during the blastocyst stage of development when the cells are

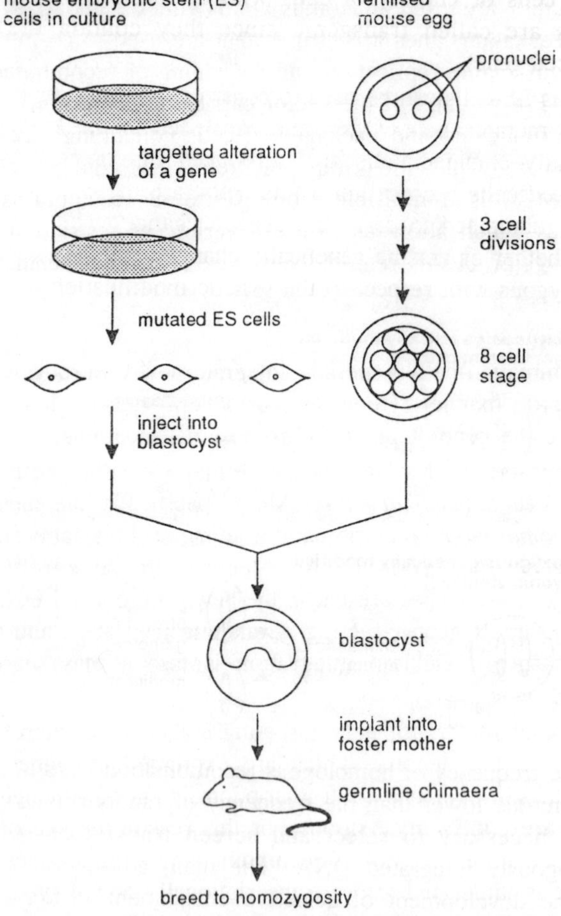

Fig. 16.6. Gene targeting in ES cells to give homozygous mice.

still pluripotent. Cells can be isolated and grown in culture for many generations and are of particular interest because they can be manipulated in culture and then re-introduced into mouse embryos. The animals which grow from these embryos are said to be *chimaeric* (chimeric), since they consist of cells derived from the genetically manipulated ES-cells as well as normal cells. (The chimaera is a mythological monster consisting of the head of a lion, the body of a goat, and the tail of a serpent. In biology it has come to mean an organism that consists of at least two genetically different kinds of tissue). The next generation of animals will either consist entirely of normal cells or entirely of cells with the desired change (and these animals are called transgenic since they contain the artificially introduced gene).

Thus ES-cells can be used to construct a whole organism with the specific mutant allele. Often mice are used in these experiments. If genetically changed cells are introduced into a blastocyst and the introduced cells become germ line cells, are the offspring of such an animal guaranteed to carry the genetic change? The answer depends upon whether or not the genetically changed cells are homozygous or heterozygous with respect to the genetic modification.

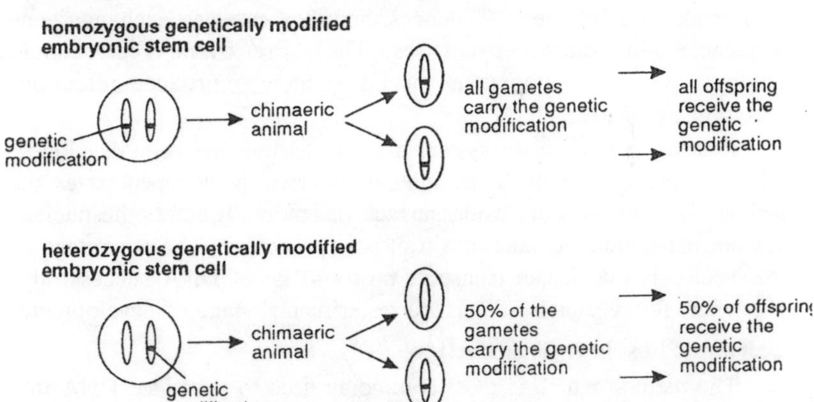

The frequency of homologous recombination is still several orders of magnitude lower than the frequency of random integration so it is always necessary to select and screen transfected cells lines for homologously integrated DNA. The main advance in this area has been the development of selective enrichment strategies and rapid screening schemes using PCR (the polymerase chain reaction).

Introduction of the Recombinant DNA

Retroviral Infection

Retroviruses were obvious candidates for use as vectors since the DNA of retroviruses becomes incorporated into the host cell genome. Many useful vectors have been derived from retroviruses, in particular avian retroviruses such as *Rous sarcoma virus* (RSV) and murine retroviruses such as Moloney murine leukaemia virus (M-MuLV).

In the retroviral vectors, the gene of interest replaces some of the viral genes which are needed for replication so the recombinant vector has to be introduced into a very special cell line which provides the missing retroviral functions (called *packaging cell lines*). The engineered retroviruses are produced in the packaging cell lines and the secreted (defective) virus from these cells is used to infect the target cells, where integration and expression occur efficiently, without the risk of producing infectious virus.

Despite their efficiency at introducing DNA, viral vectors have a number of disadvantages. Viral infection sometimes changes the cell's properties, for example epithelial cells may lose their cell surface polarity. In addition, certain retroviruses are acutely oncogenic which precludes their use in a number of types of work. Retroviruses are also prone to delete gene sequences, and they can also exchange gene sequences with other retroviruses. Thus a retroviral vector could combine with an endogenous viral sequence to produce infectious recombinant viruses.

Recently a new vector system based on viruses has been developed. DNA is attached onto the outside of the virus which penetrates the cell via its normal route using surface receptors. It enters the nucleus via normal cellular uptake and transport processes. The advantage of this vector is that it can transport up to 48 kb of DNA successfully. However, this vector is still in the experimental stage of development.

Calcium Phosphate Transfection

The method which is most commonly used to introduce DNA into cells, is based on co-precipitation of DNA with calcium phosphate first described for the infection of cells' with adenovirus 5 by van der Eb and for the transfer of cellular genes by Wigler. This method is more commonly known as *transfection*. The DNA is dissolved in a buffer containing calcium chloride, and an equal volume of buffer containing phosphate is added. A DNA precipitate is allowed to form at room temperature and this is added directly to the cells. The cells are incubated with DNA-calcium phosphate co-precipitate overnight and the cells are then placed in selective media to isolate stable cell lines.

It is not entirely clear what the function of the calcium is, although it probably concentrates DNA on the cell membrane to further facilitate uptake of DNA. In addition, calcium protects DNA from hydrolytic nucleases. The co-precipitate sediments onto cells and becomes adsorbed onto the cell membrane, where it is taken up by (phagocytosed by) the cells. A small proportion of the DNA becomes stably integrated into the nuclear genome.

Calcium phosphate transfection is still frequently used for mammalian systems where possible. However, some cells (for example bone marrow cells) do not like having the solid calcium phosphate precipitate adhering to them, and some cells are simply hard to transfect using this method. There are several other techniques for introducing DNA into these cells.

Protoplast Fusion

Another approach toward achieving high efficiency gene transfer has been to use direct fusion of cells to bacterial protoplasts. The bacterial cell wall is removed using lysozyme, and the bacterial protoplast is fused to the eukaryotic cell using polyethylene glycol. The main advantages of this method are that you do not have to purify the DNA from bacteria before introducing it into animal cells, and that the uptake of DNA is much more efficient, with a success rate of up to 100%.

Lipofection

Another method of gene transfer which relies on the fusion of membrane systems is *lipofection*. DNA in solution complexes efficiently with cationic lipids which form a single bilayer vesicle around the DNA called a *liposome*. When these liposome vesicles are added to eukaryotic cells, they seem to fuse with the plasma membrane and the DNA is taken up by the cell.

Electroporation

The cell membrane can be made permeable by exposing cells to a brief electric shock of several thousand volts. High voltage electric pulses make transient holes in the plasma membrane and any DNA which is present in the cell suspension can be taken up through these holes. The efficiency of this method may be very high. The exact conditions for electroporation, however, vary for different cell types so a lot a preliminary experiments have to be done to optimize the conditions. Thus this method is usually used only for cells which cannot take up DNA by transfection, such as myeloma cells.

Microinjection

Yet another method of introducing DNA into cells is by microinjection. Small volumes of DNA (10-100 fl, that is 10^{-8} to 10^{-7} μl) can be microinjected into the nucleus of tissue culture cells using a glass micropipette with the aid of a micromanipulator. Amphibian oocytes are popular targets for this method because of their large nuclei, but this method is not used very often for mammalian cells since it requires expensive and sophisticated equipment and considerable expertise. Obviously, cells have to be microinjected one at a time, and this is a major disadvantage of this method.

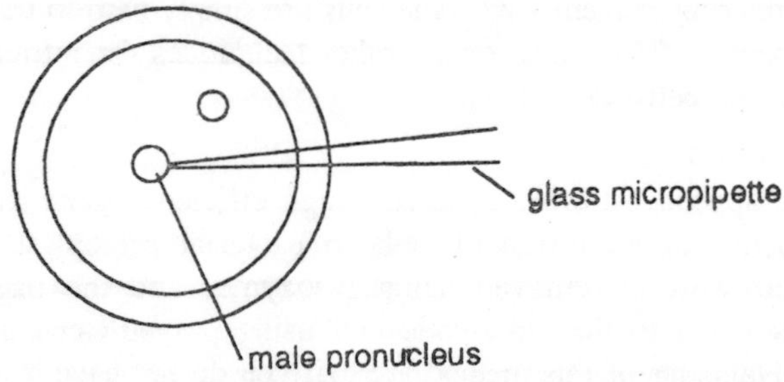

Fig. 16.7. Diagram to show the microinjection of DNA into fertilized mouse oocytes using a glass micropipette.

Microprojectiles (Biolistics)

The most novel mechanism of introducing DNA into cells in culture is called *biolistics* (which stands for biological ballistics), and this method is used especially when the more standard techniques do not work for a particular cell line. The method uses pressurized helium to shoot tungsten or gold particles coated with DNA into the cells. Although this method is still in its infancy, it is likely to become an important mechanism of transfection.

Concatemer Formation and Selection

When cells take up foreign DNA, whether by calcium phosphate precipitation, electroporation, or microinjection, the DNA seems to form concatemers before it is integrated into the genome.

We remind you that concatemers are linearly repeated units of DNA. We can represent this process thus:

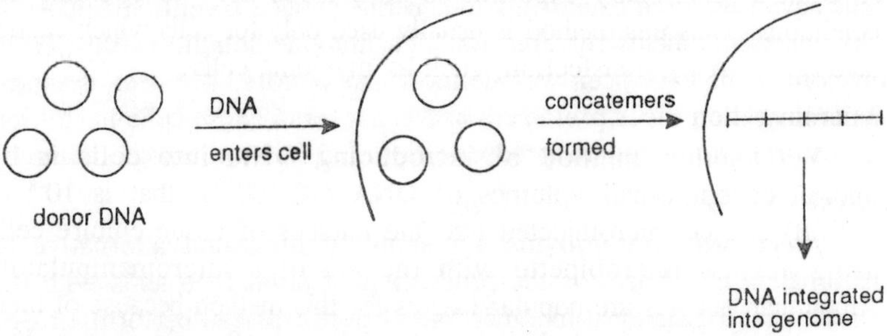

If a mixture of plasmids is used heterogenous concatemers would be produced. Thus:

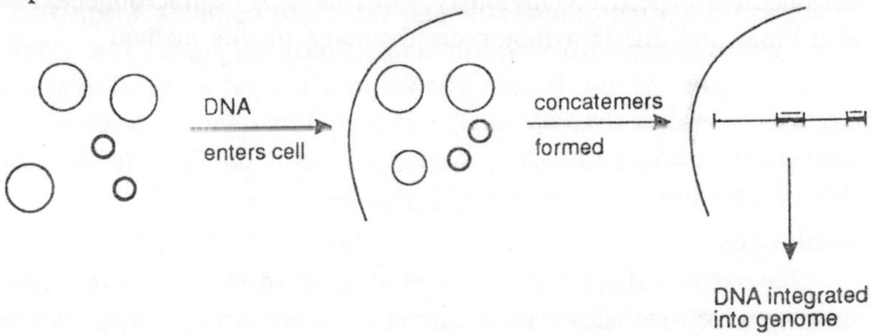

Thus selectable marker genes do not have to be included in the expression vector but can be added separately during the introduction of the genetic material. The transfection of the DNA of interest plus genes conferring resistance to antibiotics is known as co-transfection. An excess of the gene of interest is used so that most of the cells expressing the selectable marker will also be expressing the gene of interest. A typical co-transfection experiment would use 20 μg plasmid carrying the gene of interest and 1 μg of the resistance plasmid. Two days after transfection cells are split 1:5 in selective medium containing either 1 mg/ml G418 or 200 ng/ml hygromycin B. Visible colonies are isolated after 2 weeks, expanded, and tested for the presence and expression of the desired construct.

Host System

Myeloma Cells

Myeloma cells are the malignant counterparts of immunoglobulin secreting plasma cells. These cells are specialized at secreting and they have the machinery to add and modify sugar residues such as those found in mammalian glycoproteins. For many years, myeloma cells have been used extensively to generate antigen specific hybridomas. The genetic elements required for high immunoglobulin expression in myeloma cells have been well defined, and vectors have been designed containing the proper promoters and enhancers. These cells have been the natural choice for the production of modified immunoglobulins.

Insect Cell Lines

Insect cell lines provide a potentially important alternative to mammalian cell lines for the production of animal proteins with the correct post-translational modifications. Insect cells can perform a range of post-translational modifications which confer the correct folding, biological activity, and antigenicity on expressed proteins. Glycosylation occurs at the correct sites, although the more complex forms of N-linked glycosylation involving trimming of terminal sugars and addition of other sugars are not found. Thanks to a natural expression system they can provide extremely high yields of recombinant protein. One particularly useful expression system is based on the baculoviruses which infect insect cells but not vertebrate cells.

Baculovirus

The baculoviruses have a large double stranded circular DNA genome, which includes a gene coding for a protein called *polyhedrin*. This protein represents upto 50% of the total protein in the cell. During normal infection, the viruses produce nuclear inclusion bodies which consist of virus particles embedded in a matrix of polyhedrin. Transcription of the polyhedrin gene is driven by an extremely active promoter which is ideally suited for driving the expression of foreign genes. The gene of interest can be inserted just downstream of the polyhedron promoter, since the polyhedrin gene products are dispensable.

A large number of foreign genes from bacteria, plants, and animals have been expressed using baculoviruses and a number of baculovirus vectors have been constructed. High levels of expression can be obtained upon infection of cultured insect cells with these recombinant baculovirus vectors. A popular cell culture system uses cell lines derived from the fall armyworm, *Spodoptera frugiperda*. Sf9 cells can be cultured at 20-28°C in insect cell culture media which are commercially available. Upon infection of Sf9 cells with the recombinant plasmids high levels of expression of the foreign gene are achieved.

CHARACTERIZATION OF TRANSFECTED CELLS

We have covered a number of methods of introducing DNA into animal cells, although some of the more commonly used methods are not very efficient. It is therefore essential to identify those cells which have stably incorporated the foreign DNA and are able to express it. One technique to examine whether specific DNA sequences are present is *Southern blotting*.

The most common way this is done is to extract the DNA from the test material, cut it with a restriction enzyme and separate the fragments using gel electrophoresis. The separated fragments are then 'blotted' onto a membrane and melted to form single stranded DNA. These are then challenged with labelled DNA probes. If sequences complementary to those of the probe are present in the blotted DNA fragments, then the probe will form hydrogen bonds with the sequence-bearing fragments.

The presence of particular RNA sequences in cells can be assayed in a manner similar to that used for DNA except that in this case it is called *Northern blotting*.

RNA is extracted from the cell and separated according to size by gel electrophoresis. Then this RNA is blotted onto a membrane which is then challenged with labelled probes which will hybridize with the appropriate RNA molecules. You should note that this is not the only technique available for detecting the presence of specific RNA molecules. The development of polymerization chain reaction (PCR) techniques provides a method for making multiple copies of nucleotide sequences *in vitro*. This technique is, therefore, valuable in enabling us to produce copies of sequences either to provide sufficient test material in hybridization studies or to produce the radioactive probes.

The detection of specific protein either depends upon measuring a particular activity of the protein (e.g. enzymatic activity) or its ability to bind particular antibodies. In this latter case, the cellular proteins are usually separated by electrophoresis and blotted onto a membrane. Then these membranes are challenged with labelled antibodies. Often the labels used are fluorescent dyes (hence immunofluorescence) or they may be tagged with radioactive isotopes. This technique of electrophoretically separating proteins, blotting them and allowing them to react with antibodies is called *immunoblotting* or *Western blotting*.

Medical Applications of Genetic Engineering

In this chapter we have concentrated on the biotechnological applications of genetic engineering. However it is becoming more and more likely that genetic engineering will soon play an important therapeutic role in medicine.

Genetic Disorders

Ultimately, it will be possible to correct genetic disorders by replacing defective genes. Certain types of genetic disease will be particularly suited to treatment by gene therapy in the near future. These are the single gene disorders where the gene is not extremely stringently regulated. Two types of gene therapy are envisaged:

- Germline gene therapy;
- Somatic cell therapy.

Some somatic cells are particularly suitable to act as recipients for genes. The haemopoietic system has pluripotent stem cells, and, in animal models, these cells have been infected with recombinant retroviruses carrying a human gene. The gene was found to be expressed in the peripheral blood lymphocytes for over 10 months, although expression was not very efficient.

Advances have also been made in the introduction of foreign genes into cultured skin and blood vessel cells by insertional mutagenesis.

Retroviral-mediated gene transfer is the current method of choice for clinical studies, although any treatment involving retroviruses carries a risk that the defective retrovirus may recombine with non-defective retroviruses, which would allow its spread to inappropriate sites. In 1989, the first approved human gene experiment was conducted in the United States. Tumour infiltrating lymphocytes (TILs) were engineered to carry a marker gene for neomycin resistance into tumours of patients with terminal melanoma. In 1990 the same research group transfused a small girl suffering from adenosine deaminase deficiency with lymphocytes which had been infected with a retrovirus bearing the adenosine deaminase (ADA) gene. The trial appears to have been successful, although this is obviously too small a sample size to draw any statistically valid conclusions.

Cancer treatment

Another application of gene therapy is in the treatment of localized proliferating tumours. In an animal model, rats with cerebral glioma were given an intra-tumoural injection of mouse fibroblasts that were producing a retroviral vector expressing Herpes simplex virus thymidine kinase. The thymidine kinase product turns any cell which produces it into a target for the anti-viral drug ganciclovir. After 5 days, during which the retroviruses transduced the neighbouring proliferating glioma cells, the rats were treated with the anti-viral drug. The gliomas regressed completely in 11 out of 14 rats, and in none of the controls.

The use of a similar system in inoperable human brain tumours was approved by the National Institute of Health in the United States in June 1992.

Delivery of products in vivo

Many humans diseases could be cured if the right product could be made by a cell at the right time. One of the directions of clinical research has been to move from the production of the missing compounds by pharmaceutical companies, towards engineering patient's cells to make the products themselves. The obvious candidates for genetic engineering are those cells which can be easily taken out of the body and replaced after they have been modified *in vitro*. There has been much interest in blood and bone marrow, although these are difficult to infect with retroviral constructs, and, when these cells mature, expression of the recombinant protein stops. Some scientists have turned to endothelial cells (the cells which line the bloods vessels and airways of the lungs) which are easier to infect. After introducing the gene of interest into endothelial cells, the cells are re-introduced into the animal in the form of organoids. Such organoids have been re-implanted into the peritoneal cavity of rats, where they induced the formation of new blood vessels.

INDEX